Nellita
July, 1990

To Be and Not To Be

To Be and Not To Be
Negation and Metadrama in *Hamlet*

James L. Calderwood

New York
Columbia University Press
1983

Library of Congress Cataloging in Publication Data

Calderwood, James L.
To be and not to be.

Includes bibliographical references and index.
1. Shakespeare, William, 1564–1616. Hamlet.
2. Negation (Logic) in literature. I. Title.
II. Title: Metadrama in Hamlet.
PR2807.C24 1983 822.3'3 82-22214
ISBN 0-231-05628-1
ISBN 0-231-05629-X (pbk.)

Columbia University Press
New York Guildford, Surrey

Clothbound editions of Columbia University Press books are Smyth-sewn and printed on permanent and durable acid-free paper.

For Stuart and Ian

To thee no star be dark

CONTENTS

PREFACE

PART OF the first chapter of this book appeared as an article called "*Hamlet*: The Name of Action" in *Modern Language Quarterly* (December 1978); I am grateful to William H. Matchett, editor and good friend, for permission to reprint it here in revised and enlarged form. I am also grateful to my colleagues Murray Krieger, Harold E. Toliver, and James McMichael for reading and commenting on various parts of the manuscript, and to William P. Germano, Assistant Executive Editor at Columbia University Press, for ushering it so expeditiously into print.

I should perhaps apologize for the unusual length of the chapters, which seem to expand exponentially, like men in buckram, as the book proceeds and my plight grows more desperate. To render the turns of argument somewhat clearer I have supplied subtitles to the different sections and have periodically glanced backward, not merely to see, as the late Satchel Paige would say, if something were gaining on me, but to discover where I have been and to chart my location for the reader. For convenience I have often referred to familiar scenes and to episodes within scenes descriptively, following the practice of Harry Levin in *The Question of Hamlet* (New York: Oxford University Press, 1959). The descriptive title and lineation of these "scenes" is as follows, according to the text I am using— that of David Bevington in *The Complete Works of Shakespeare*, 3d edition (Glenview, Ill.: Scott, Foresman, 1980):

Council Scene	1.2.1–128
Nunnery Scene	3.1.89–191

Play Scene	3.2.89–293
Prayer Scene	3.3.36–98
Closet Scene	3.4.1–224
Graveyard Scene	5.1.1–299
Duel Scene	5.2.1–405

And since critics sometimes disagree about which of Hamlet's speeches are soliloquies, let me record my line-up here:

First Soliloquy	1.2.129–159
Second Soliloquy	1.5.93–113
Third Soliloquy	2.2.549–606
Fourth Soliloquy	3.1.57–89
Fifth Soliloquy	3.2.387–398
Sixth Soliloquy	3.3.73–96
Seventh Soliloquy	4.4.32–66

J.L.C.

Laguna Beach, California

Prologue

COMING UPON what he calls Death's feast of princes at the end of *Hamlet*, Fortinbras asks in effect, "What has happened here?" The task of replying falls to Horatio, and rightly so. At a time when "men's minds are wild" it is fitting that the man who is more an antique Roman stoic than a Dane should effect a catharsis by telling Hamlet's story, doing for his mute audience of Danish nobles within the play what Shakespeare has done for his audience in the Globe theater. Horatio's reply to Fortinbras will transform the present-tense immediacy of tragic life into the mediated pastness of narrative, imparting point and order to what seems a chaos of death and destruction. Horatio can do this not merely because he is a stoic philosopher, one who is not passion's slave, but also because he is, as his name implies, a "speaker." Moreover, he alone can speak with authority to "the yet unknowing world / How these things came about" because he more than anyone else knows what has happened:

> So shall you hear
> Of carnal, bloody, and unnatural acts,
> Of accidental judgments, casual slaughters,
> Of deaths put on by cunning and forced cause,
> And, in this upshot, purposes mistook
> Fallen on the inventors' heads. All this can I
> Truly deliver.[1]

In one sense Horatio's story symbolizes and fuses with Shakespeare's play. The dead bodies will "high on a stage be placed to the view," and like a script Horatio's story will accompany and explain this "staging." No doubt the first words of his re-presentation will circle back to tell of the entrance of a Ghost on the castle platform, rather as the unconcluding "the" of *Finnegan's Wake* returns by a "commodious vicus of recirculation" back to the unbeginning "riverrun" of its opening. From this point of view Horatio's story has the authenticity of Shakespeare's play itself, from which it issues and into which it flows.

In another sense, however, Horatio's story is merely a bad quarto of Shakespeare's play, a pirated edition based on memorial reconstructions by an actor who, though he knows much, cannot possibly know all that has happened in the castle at Elsinore or on the stage of the Globe. Indeed, from Horatio's limited perspective and hearsay evidence, what *has* happened here? Will his story bring definition and clarity to the ghostly images in Claudius' house of mirrors? From "accidental judgments, casual slaughters," and "purposes mistook" will he extract design, causation, and fulfilled intent? Or like his namesake, the "Orator" in Ionesco's *The Chairs*, will Horatio be reduced to writing something like "Angelfood" on a slate for his baffled audience?

Of course Horatio is not the Orator of *The Chairs*, nor is Shakespeare Ionesco. Shakespeare's play proceeds from the absurd, not toward it, from a world perceived as "weary, stale, flat, and unprofitable" toward one in which a "special providence" can be seen even in the apparently random "fall of a sparrow." That movement toward meaning will, we assume, be reflected in the story Horatio tells. And yet we may take leave to wonder how much meaning and what meanings—how much of the "why" embedded in the "what" of Fortinbras' hypothetical question—will be recaptured by Horatio.

Ultimately, of course, like so many of the questions in *Hamlet*, the question of Fortinbras reaches out of Hamlet's world and Shakespeare's Globe and addresses itself to us as well. Like Prufrock, we are not by any means Prince Hamlet, nor were meant to be; but in some degree we are all meant to be Horatio pondering, as the bodies are carried up to the stage, this most difficult question asked by the rough young prince from Norway. For my own part, answering Fortinbras is too ambitious an enterprise. The questions that have caught my attention are much smaller and less stridently posed. There needs no Norwegian prince in armor to make me wonder, for instance, simply why Shakespeare has been so parsimonious in his giving of names that the hero and his dead father must both share the name of "Hamlet," even as the Norwegian prince and his father must share that of "Fortinbras." I have also wondered about Hamlet's famous delay of his revenge. Not so much *why* he delays the revenge, an issue that leads into the labyrinths of character psychology, as why he advertises his delay, an issue that leads out onto broader avenues of dramatic form and generic structure. Why, again, does Claudius poison Hamlet's father in such an unorthodox manner, through the ears? And why does this same "Claudius," who sits on the throne of Denmark, not exist as such in any theatrical presentation of Shakespeare's play? Why, on the other hand, does someone we never see, an irrelevant noncharacter by the name of "Claudio," acquire verbal presence in the play at a significant moment? Why does Horatio, whose name implies that he is an "orator," fall increasingly silent as the play proceeds? Why does Hamlet at the end of the Closet Scene begin his advice to his mother with an ostentatious double negative—"Not this, by no means, that I bid you do"—and then portray that "Not this" with repugnant vividness? And why, when he finally gets around to taking his revenge, does Hamlet kill the King twice?

These are small questions about minor issues, mere motes perhaps, and yet they trouble the mind's eye. Perhaps some of them could be dismissed if we were content to assume that Shakespeare was simply careless, or that the compositor erred, or that the text was revised. But I am reminded that the late Sigurd Burckhardt, who had a genius for detecting motes and for perceiving Blakean infinities in them, deplored such "explanations" as a shirking of critical responsibility.[2] He argued that the Book of the Play, like the medieval Book of Nature, is a single unified order, totally coherent and interpretable, and therefore infallible—not in actual fact, of course, but as a working hypothesis. As a result the critic who encounters a puzzling discrepancy is obliged to exert every effort to bring it into alignment with the total order, and, failing that, to assume that the failure lies in himself, not in the play or playwright. On this view apparent mistakes and deviations are to be regarded as challenges to our theories, clues to our errors, and opportunities for critical advancement. Geoffrey Hartman says something similar when he observes that "interpretation is like a football game. You spot a hole and you go through. But first you may have to induce that opening."[3] With his typical hospitality, however, Shakespeare often relieves us of the task of inducing our openings. He tends to employ what is called in professional football a "flex" defense, in which the linemen leave certain holes invitingly open and the linebackers close them with routine ferocity, at a great cost to ballcarrying critics.

Burckhardt's faith in the integrity and coherence of the text sounds almost antiquated in our current age of critical unbelief, when new philosophy calls all in doubt, things fall apart, and the center cannot hold, even off the football field. Discrepancies are now not merely dismissed as mistakes, as Burckhardt feared, but are made the basis for the deconstruction of coherence and integrity themselves, the decentering of structure. And Burckhardt's notion of total interpretability has given way in some quarters to a focusing on marginal features of a work, not so much to elucidate the text as to convert it into an illustrative example of a more general subject like language, narrative, or fictionality. I am not altogether innocent of these practices myself. Interpretation, for instance, in the sense of explicating the meaning of the text, is not the entire aim of this book. The first chapter presents a reasonably self-contained (not to say impenetrable) reading of *Hamlet*, and the remaining chapters are at least intermittently concerned with the meanings of various aspects of the play. But because *Hamlet* is not merely a text but a play, I have addressed myself, in addition to its meanings, to its mode of being as a staged performance, as both a verbal and visual sign. For its mode of being is the ground for its meanings.

I use the word "meaning" in its plural form advisedly, since both verbally and visually the play is dominated by Shakespeare's negative mode, which multiplies entities and meanings in a baffling manner. That is not to repeat the familiar notion that *Hamlet* is, even among Shakespearean plays, extraordinarily ironic in the sense of subverting its apparent themes and meanings by juxtaposing them with their opposites. That sort of dialectical irony—which generates an unending series of graduate papers and theses entitled "Love and Honor, or Reason and Passion, or Nature and Grace in (a play of your choice)"—one can always resolve by discovering, or, failing that, by inventing, a transcendent synthesis. But Shakespeare's negative mode is more radical and less resolvable. Its complexity is suggested by my substitution of "and" in my title for the "or" of Hamlet's famous expression. In place of Hamlet's implied and Aristotle's explicit law of the excluded middle (a thing is either A or not-A) we have Shakespeare's law of the included middle (a thing may be both A and not-A).[4] This is by no means the same thing as saying, with some of the more extreme devotees of indeterminacy, that a thing is neither A nor not-A; nor is it a license to valorize *any* A or any not-A.

At first glance *Hamlet* seems bent on dissolving all of its A's into not-A's, including itself. The play, that is, seems dedicated to its own deconstruction by virtue of its employment of negation and erasure, in its stress on silence and secrecy, in its self-conscious creation of divisions and junctures, and in its metadramatic denial of its own illusions. Much of my second and third chapters is devoted to examining this negative mode of now-you-see-it-now-you-don't, which goes a good way toward leaving us as bemused in the theater as Hamlet is in Denmark. Yet, as I argue more fully in the Epilogue, deconstruction, the emptying of value from Hamlet's world and of mimetic presence from Shakespeare's illusions, is where Shakespeare begins, not where he ends. As part of a more subtle illusion he passes, with Hamlet in his world, from deconstruction to reconstruction.

As that might suggest, my own reading of *Hamlet* is not itself deconstructive, in either of two senses. That is, first, I am not attempting, in an era of definitive revelation, to devalue the interpretations of the play offered in the past by a variety of critics. I confess I am not in sympathy with views like that of Rebecca West when she says, "It is quite certain that [Shakespeare] wished to present Hamlet as a bad man, because he twice makes him rejoice at the thought of murdering men who had not made their peace with God."[5] On the other hand, although I acknowledge that Hamlet after the sea voyage is considerably changed, I would by no means characterize him as a regenerate Christ-figure in the style of another school of critics. Instead, I am interested in exploring the kinds of dramatic indirection, duplicity, and reversibility in the play that make it not only possible but likely that critics will arrive at such apparently incompatible positions.

Second, my reading of *Hamlet* is on the whole metadramatic rather than deconstructive. Deconstruction, it seems to me, is built into the play, as it is in many literary works, to the extent that *Hamlet* repeatedly insists upon its own fictionality, or in this case theatricality, and addresses itself to the nature of dramatic illusion. But this deconstructive theme is but a part, albeit a highly significant part, of a metadramatic reading that regards the more complete subject of the play to include, not only its own theatricality, but also its own digressive structure, its self-consuming nature as a performance in time, its concrete universality, the creative negativity of its language, and its relations specifically to its source play, the *Ur-Hamlet*, and to its genre of revenge tragedy. Moreover, in dealing with these aspects of the play's modes of being and doing, I have not, I hope, lost sight of the play as an illusion of life and mimetic presence whose humor and grace and pathos take up permanent residence in our emotional experience. Meanings are made by signs, illusions by theatrical performances, and music by instruments; but the fact that the one cannot exist without the creative mediation of the other does not mean that we should substitute the means for their ends and live only with signs, theaters, and instruments. So without becoming overly megalomaniacal, without claiming with Horatio "All this can I/ Truly deliver," I have tried not to forget that question of Fortinbras, "What *has* happened here?" while formulating hypotheses about the question Shakespeare seems also to ask, "What has *not* happened here?" As my title suggests, the answer to the one is inextricable from the answer to the other.

PART I

Hamlet: The Name of Action

ONE

NAMES AND MEANINGS

OFF THE coast of Wales to the northwest of Caernarvonshire is the island of Anglesey, which the Romans (and Milton in "Lycidas") called Mona, and on the landward side of this island is a town with the name of Llanfairpwllgwyngyllgogerychwyrndrobwillllandtysiliogogogoch. As one might expect, it means "The church of St. Mary on the pool of the white hazel by the raging whirlpool near the church of St. Tysilio of the Red Cave." One wonders of course why the local folk, who delight in baffling visitors with the pronunciation, should have burdened themselves originally with such a backpack of sound. My first guess imputed to the name-givers a kind of scientific or cartographic motive. Wanting to get the town exactly situated, I assumed, they invented a name embodying a set of landmarks, a name as much like a map as possible, so that by standing by the pool with the white hazel near the raging whirlpool and the church of St. Tysilio with its red cave, and consulting the name, which you would unfurl like a scroll, you would know that you were indeed at the right place. Thus the name seems to act as a wonderful particularizer, announcing "This place here and nowhere else."

On the other hand, if the name did issue from a cartographic impulse, it is rather too self-referential to serve as a map. In fact it is a closed circle of designation: you can locate not merely the church of St. Mary but any of the defining terms simply by turning the name this way or that. But unless you already knew the location of Llanfair P.G., as the local post office now calls it, you could not find it by means of the name.

The name is a self-enclosed system of mutually defining places, rather like a dictionary that sends us from one term to another—from, say, "being" to "actuality" to "existence" and back to "being"—in order to understand the first word looked up. Its admirable particularity thus deprives the name of meaning by disconnecting it from frames of geographical reference.

If we glance at the flyleaf of Stephen Dedalus's geography book, we shall see that the cartographic spirit works with greater rigor in Ireland than in Anglesey:

Stephen Dedalus
Class of Elements
Clongowes Wood College
Sallins
County Kildare
Ireland
Europe
The World
The Universe

Stephen's personal chain of being works upward from his particular case through ever broadening classes toward the universe and God as the class of all classes. Going up gets Stephen to God's address, and going down gets God to Stephen, who does not want God to confuse this particular Irish sparrow with any of the others. By situating his name within such a snug hierarchy of classes, Stephen escapes loneliness by acquiring the company of larger wholes. By contrast, Llanfair P.G. remains uniquely isolated, comforted only by its own gregarious syllables. One suspects, then, that the marvelous name must have been bestowed by poets instead of cartographers, since far from revealing the location of the town, the name conceals it within a song of self-celebration.

With these preambling remarks I seem in danger of not finding Shakespeare and *Hamlet*, let alone Llanfair P.G. There is a route from the one to the other, however, for the hero of *Hamlet* appears by virtue of his proper name to acquire something of the uniqueness and isolation of the Welsh island town. As there is beyond doubt but one Llanfairpwllgwyngyllgogerychwyrndrobwillllandtysiliogogogoch, certainly there is but one Hamlet. Or is there? After all, if I were to say aloud,

"What do you think of Hamlet?" you might well wonder if I meant Hamlet the hero of the play, Hamlet the father of the hero of the play, or *Hamlet* the play itself. Before settling down with the theatrical issue, let me wander a bit further by talking about names in general, with a reassurance that we will, with all of this verbal triangulation, actually come upon Shakespeare's play in due course.

As a class, proper names are the linguistic ultimates—the verbal quarks and neutrinos—of particularizing, the point at which an existentialist reduction would have to stop, since it is at that point that meaning is stripped from words and we are left to confront sheer being. Thus Wittgenstein observed that "a name cannot be dissected any further by means of definition: it is a primitive sign";[1] and Gilbert Ryle adds that "dictionaries do not tell us what proper names mean—for the simple reason that they do not mean anything."[2] Instead of meanings, proper names are said to have "bearers"—the horse called Bucephalus, the ship called *Queen Mary*, the philosopher called Gilbert Ryle. "Proper names are arbitrary bestowals," Ryle says, "and convey nothing true and nothing false, for they convey nothing at all."[3]

Meaning, like men, is begotten by coupling. In nature each bird and bush exists as itself, unlinked to any class: it simply is. The bearer of a proper name is also simply there, a "this" to which we may point. Only when we couple the perceptual object with a verbal category by saying, for instance, "This is a bird" does "being" graduate into "meaning." To say "This is Bucephalus," however, is no more than to say "This is this"—or, with Feste, "That that is is." Radical thisness acquires meaning by merging with a class, by uniting, like Stephen Dedalus, with a larger whole.

Not all proper names are alike, though. At one extreme we find those that particularlize almost as impressively as Llanfair P.G. For instance, Oscar Wilde was born Oscar Fingal O'Flahertie Wills Wilde. Such precise differentiation suggests that Dublin at the time of the writer's birth must have been teeming with Wildes. If so, perhaps it is understandable that as this Wilde grew older he jettisoned most of his names and hoped eventually to be known simply as "The Oscar" or "The Wilde"—names designed not to distinguish him from others within the same class but to proclaim him as a class in himself.

At the other extreme are what might be called "common" proper names. In his essay "Of Names," for instance, Montaigne points out that "history has known three Socrateses, five Platos, eight Aristotles, seven Xenophons, twenty Demetriuses, twenty Theodores; and just guess how many it has not known."[4] Here the proper name, grown common, no longer particularizes. Of course no one wonders who is meant if we say Socrates, Plato, or Aristotle. But it was the class or universal aspect of the name that appealed to the parents of the third Socrates or the fifth Plato, who no doubt hoped for a generic transfusion of distinction from the original bearer of the name.

Between these extremes are the proper names normally bestowed on people: Gilbert Ryle, Ludwig Wittgenstein, and so forth. And here, it would seem, contrary to Gilbert Ryle, proper names do have a certain meaning and are not arbitrary bestowals. The parents of Nick Bottom in *A Midsummer Night's Dream*, for instance, might have given their son, as freely as they liked, the name of John or Fred or Nick, quite in keeping with Ryle's claims. But they were not free to call him Nick Quince or Nick Flute or Nick Stout, simply because his surname represents a class that is not arbitrary. As it is, the forename "Nick" particularizes the surname "Bottom" by singling out this unique member of the Bottom family—the only one, presumably, whose bottom extends to his top, to the delight of fairy queens. Or, the other way round, the surname "Bottom" universalizes the particular "Nick" by incorporating him into a larger class. Together, the names comprise a concrete universal abbreviating the descriptive "Nick is a Bottom." The individual Nick, who simply is, acquires meaning quite literally through his "relations," through his membership in a family. By the same token, the Romeo who stands stricken in the orchard beneath Juliet's balcony exhibits a roselike, nominalistic uniqueness. By whatever word we or Juliet call him, he is simply and unrelatedly there, like Adam in an earlier garden. But add to "Romeo" the surname "Montague" and he becomes most dangerously meaningful as a member of a feuding class.

TWO

NAMES AND IDENTITIES

FINALLY, WHAT about names in *Hamlet?* First, this is a play in which Shakespeare the dramatist godfather has been both careless and careful with his name-givings. On the one hand his Danes are by no means recognizably Danish as far as their names are concerned. Harry Levin says, "If Marcellus and Claudius are Latin, Bernardo and Horatio are Italian, and Fortinbras signifies 'strong arm' not in Norwegian but in French (*fort-en-bras*)."[5] Scandinavian names like the Germanic Gertrude are hard to find. But if this suggests carelessness, a certain carefulness is implied by the characterization of the contentious Fortinbras as "strong arm" and of Horatio, the scholar who addresses the Ghost and who is to deliver Hamlet's story to the surviving audience, as "orator."

However, Shakespeare seems less concerned with the nationality of his names than with their concrete universality, or lack of it. The play presents us with one Ophelia, one Gertrude, one Laertes, one Polonius, one Horatio, one Osric, and so on. That these characters require only a forename suggests their singularity within the play but not, unlike such "forenamed" personages as Alexander or Cleopatra, their individuality, for they certainly have their generic aspects. Osric, Horatio, and Polonius, for instance, are clearly types of the fop, the faithful friend, and the foolish counsellor. But although they may be linked as types with other characters in other plays and works of literature, within *Hamlet* itself they are distinctly singular, disconnected in nature and in name from the other characters, except of course for Rosencrantz and Guildenstern, whom not even names can distinguish.

Claudius too has but one name, and yet his case is different—even puzzlingly different. That is because, as "Claudius," he does not exist in any performance of Shakespeare's play. We call him Claudius because that is how he appears in the list of *dramatis personae,* which is based on the original stage direction to Act 1, Scene 2, informing us of the entrance of "Claudius, King of Denmark." Audiences in the theater, however, do not hear stage directions or, in Shakespeare's time, receive lists of *dramatis personae.* Nor do they hear the name "Claudius" spoken by any of the characters in the play. "Claudius" is without exception referred to as "King" (37 times), "Denmark" (5 times), or "Majesty" (8 times). This is puzzling on two counts: first, because in the stories told by both Saxo Grammaticus and Belleforest the usurping king had a name (Feng or Fengon), and second, because each of Shakespeare's other kings (Richard II and III, the Henrys, John, Macbeth, even the regal Lear) is called by his individual name as well as by his title. Only Claudius is nameless apart from his royal office. Why, one wonders, should it be so particular with him?

"Particular" is precisely what it is not. The particular Claudius is nominally a mystery, a blank. He exists only as the King, and to be called "the King" is by no means the same as to be called "the Wilde," since the royal class name obscures whereas the proper name proclaims individuality. Of course the character Claudius is individualized by private thoughts and feelings to which we are made privy, as in his efforts at prayer. But in titling him always "the King" Shakespeare seems to suggest a certain blurring of the individual by his regal role. Thus in Claudius' first scene (1.2) his language runs to the bland impersonality of political rhetoric, and the pronouns of corporate royalty—"we," "our," and "us"—come forth with practiced ease as he ticks off the items of state business. He himself speaks of "the King's rouse" and of himself as "Denmark" (1.2.125–127). Of course royalty is by its nature corporate, and the king the embodiment of the state, as Rosencrantz says in delivering the official Renaissance doctrine in sycophantic accents:

The cease of majesty
Dies not alone, but like a gulf doth draw
What's near it with it; or it is a massy wheel
Fixed on the summit of the highest mount,

> To whose huge spokes ten thousand lesser things
> Are mortised and adjoined, which, when it falls,
> Each small annexment, petty consequence,
> Attends the boisterous ruin. Never alone
> Did the King sigh, but with a general groan. (3.3.15–23)

In Claudius' case, however, this enlargement of the King's identity takes a special toll, in part perhaps because of the nature of his usurpation. For in seizing Old Hamlet's possessions—divesting him of life, wife, and crown—Claudius confiscates also his identity (on the principle that you are what you own). By so completely assuming the roles played by his brother, Claudius cashes in his own identity. He is now uncle-father to young Hamlet, brother-husband to Gertrude, and usurper King of Denmark. This erasure of personal identity combined with the generalizing effect of royal office reduce Claudius to a kind of cipher, as Hamlet indicates:

> The King is a thing—
> *Guildenstern.* A thing, my lord!
> *Hamlet.* Of nothing. (4.2.29–31)

If saying "This is Claudius" is, in the way of proper names, no more meaningful than saying "This is this," then at the other extreme saying "This is the King" is no more "being-ful" than saying something like "Beauty is truth." In his royal universality Claudius is replete with meaning but devoid of identity, a "thing" or "no-thing," the price of office having been paid in the coin of self.

Thus generalized, the King is well equipped to serve as "mighty opposite" to Hamlet, himself perhaps the most complex, unique, and inward-dwelling of literary characters. This uniqueness of Hamlet would seem to be reflected by the singularity of his name, as he, like the others, has but one name. Unlike the others, however, Hamlet shares his name with his dead father, Old Hamlet, he who "o'ercame Fortinbras . . . the very day that young Hamlet was born" and the Gravemaker came into his trade (5.1.144 ff.). Thus the Prince's name both is and is not his own. Insofar as it *is* his own, he is distinctively himself. Insofar as it is *not* his own, however, he yields his personal identity to family relations in the manner of Romeo when "Montague" is added. As unique "self" and generic "son" Hamlet has two dimensions of identity compressed into a single name so that he is neither wholly himself nor entirely consubstantial with his father. In a play replete with puns, even the word "Hamlet" is multivocal and resistant of easy definition.

How, then, does Hamlet define himself? Normally we define men as we define words, putting like with like, then winnowing out the distinctively unlike. Abraham Fraunce could have showed Hamlet how to go about it:

> A Definition is that which declareth what a thing is: it consisteth of two parts, the generall and the difference. Whereof the first is common to the thing defined, and all his other fellow specials, but the difference is proper onely to the thing defined, and distinguisheth it from all other his fellow specials . . . as, A man is a sensible creature endued with reason, where sensible creature is the generall, and endued with reason is the difference.[6]

As with the word "man," so with man himself. Submerged in participative tribalism, primitive man, we are told, had little realization of personal identity: "I" and "we" were largely interchangeable. A Maori, speaking of a tribal battle several hundred years earlier, will say "I defeated the enemy there," or, indicating thousands of acres of tribal land, will say "This is my land." Even Homeric man, enclosed in the *paideia* of an exclusively oral culture, was not an independent self but an inseparable part of all he had seen and heard and remembered.[7] Only gradually does such primitive unity yield to multiplicity, the tribal "generall" subdividing into autonomously different individuals in the approved Abraham Fraunce manner. Similarly, within the family, modern or ancient, only gradually does the child disengage himself from the parental matrix and acquire a sense of personal identity.

Hamlet's "problem" (to adopt the usual phrasing) is in part a matter of self-definition, inasmuch as he like all young men must distinguish himself from his father, from whom he is genetically descended and to whom he is therefore generically related. These genetic and generic aspects of identity coalesce in Hamlet's inherited name. But that of course is not all Hamlet inherits. With the return of the Ghost to tell its story and to demand revenge, Hamlet also inherits an act of filial obligation. When he swears to avenge his father, he swears in effect to relinquish his personal identity and to unite with his father not merely in name but in actional fact. That is, to adopt his father's cause—to make his father's enemy his own enemy, to assume his father's motives, goals, and pains—is to adopt his father's identity.[8] Deputies are always in some degree extensions of the authority they represent. All the more so when the deputy is a son endowed with his father's name and sworn to go about his father's vengeful business.

Hamlet, however, does not go about his father's business in very great haste, and that—his delay or irresolution—has been a notoriously vexed issue in *Hamlet* criticism. Before vexing it even further, let me succumb to Hamlet's malady myself a bit by turning to two other characters who have a rather similar problem.

THREE

Fortinbras as Self and Son

LET US look first at Fortinbras, who is something of a puzzle. Criticism has had little to say about him, and that is understandable. He does not appear, by his own or any other name, in Saxo or Belleforest; he shows up in person just once before his strident entrance at the end of our play; and he is referred to only four times. Moreover, he speaks even fewer lines than Osric—and the bibliography of Osric criticism is rather less than voluminous. Nevertheless, there is Fortinbras, moving silently at the edges of the play, raising armies against Denmark, employing them against Poland, appearing suddenly amid the carnage of the final scene to receive Hamlet's dying vote, taking charge of Hamlet's funeral, and preparing to assert his title to the Danish throne. So elusive and apparently irrelevant is Fortinbras that one wonders why he appears at all.

Another curious feature of Fortinbras is that he is the one character in the play besides Hamlet whose name is singular and identical with his father's. If this suggests additional parallels between the two princes, they are not far to seek. Like Hamlet's, Fortinbras' father has been killed—by Hamlet's father—and his father's brother has assumed the throne of Norway, as Claudius has done in Denmark.[9] Fortinbras is prompted, though not by a Ghost, to revenge, to regain the lands forfeited by his father upon his defeat by Old Hamlet. Unlike Hamlet, Fortinbras is reasonably swift to act, once he is old enough to do so. Despite the fact that the encounter between the two kings was a model of chivalric combat, its compact of terms "well ratified by law and heraldry," young Fortinbras has "sharked up a list of lawless resolutes" and seeks to recover the land "so by his father lost" (1.1.87 ff.). Not inheriting his father's crown (was he simply too young, or did old Norway pop in between the election and his hopes?) Fortinbras now seeks by conquest to inherit his land. By defeating the Danes and recovering the lost land, he would engage in a restorative act that would, as it were, annul his father's defeat and make the nominal identity of father and son symbolic of a real identity.

Thus in taking up his father's cause, as the Ghost implores Hamlet to take up *his*, Fortinbras "becomes" his father. But this fusion of identity is short-lived. If Old Fortinbras survives for a time in his son, he is killed off at last by the intervention of Claudius, who warns old Norway of his nephew's military ambitions. Distressed by the deception, old Norway, as the ambassador Voltimand relates,

> sends out arrests
> On Fortinbras, which he, in brief, obeys,
> Receives rebuke from Norway, and in fine
> Makes vow before his uncle never more
> To give the assay of arms against [Denmark].
> Whereon old Norway, overcome with joy,
> Gives him three score thousand crowns in annual fee,
> And his commission to employ those soldiers,
> So levied as before, against the Polack. (2.2.67–75)

Fortinbras then ushers his army peacefully through Denmark en route to Poland, where he is victorious, and returns to Denmark in time to hear Horatio's account of the tragedy to which he has seemed so peripheral.

One answer to the question about why Fortinbras appears in *Hamlet* is that he presents us with a clearly articulated paradigm of self-definition—of the process by which the individual emancipates himself from the bonds of family, especially in this case the father. The steps are clearly marked. First, assuming the cause of Old Fortinbras, the son identifies himself with his father. Then prompted by Norway, he forswears that cause, in effect thereby burying his dead father and giving birth to himself as an individual. Receiving authority from Norway, he takes command of an army, an act that suggests the kind of self-mastery symbolized in the stories of Joseph Conrad by one's taking command of a ship. Finally, his military successes in Poland confirm the emergence in him of an identity discrete from that of his father and distinguished in itself. He now becomes qualified to assume charge, not, oddly enough, of Norway, but of Denmark. And why Denmark? Because by indirections, it seems, Fortinbras has found directions out. The Danish lands which he sought through direct conquest to regain in his father's name have come to him after all, by way of a plot of ground in Poland so small that "the numbers cannot try the cause" (4.4.63). Self-fulfillment, it turns out, is not inconsistent with filial obligation. The two lines of self-definition—Abraham Fraunce's "generall" and "difference"—converge at the same point.

FOUR

Laertes as Son

A MORE obvious parallel to Hamlet (certainly a more frequently noted one) is Laertes, of whom even Hamlet says, "For by the image of my cause I see/ The portraiture of his" (5.2.77–78). Like Fortinbras, Laertes makes occasions inform against Hamlet, for though both have a father killed and ample excitements of reason and blood, yet Laertes acts with an impetuous urgency that is quite foreign to Hamlet. In that respect Laertes is like Fortinbras. But he is unlike Fortinbras too. For while Fortinbras first embraces and then forswears his father's cause, Laertes perseveres to the death in revenging Polonius.

Laertes' values tend toward the universal, toward "relations" and public symbols, from the start. Admonishing Ophelia against too much hope of Hamlet's genuine love, he claims that the Prince's

> will is not his own
> For he himself is subject to his birth.
> He may not, as unvalued persons do,
> Carve for himself . . . (1.3.17–20)

This, despite the fact that at Ophelia's grave the Queen herself says,

> I hoped thou shouldst have been my Hamlet's wife.
> I thought thy bride-bed to have decked, sweet maid,
> And not have strewed thy grave. (5.1.244–246)

Laertes cannot see the man who loves Ophelia for the prince obedient to his public role.[10] By the same token Laertes cannot in his own life elude the parental shadow of Polonius, and in fact is far more "subject to *his* birth" than Hamlet. Although he is advised "to thine own self be true," the very scene in which the advice is given—with Polonius conferring his blessing and imposing patterns of prudential wisdom on the departing Laertes—establishes the dominance of father over son. Once Laertes is gone, control through precept is reinforced by control through spying, as we see in Act 2, Scene 1, where Polonius coaches Reynaldo in the subtleties of surveillance. Even in distant risqué Paris, Laertes, far more than Hamlet, will be in the punning sense "too much in the sun."

Thus it is in keeping with what we know of Laertes that upon hearing of the death of his father he should forsake his life in Paris, rush back to Denmark, and address himself to sudden vengeance. The extremity of his self-abnegation is suggested by the extremity of his rhetoric:

> To hell, allegiance! Vows, to the blackest devil!
> Conscience and grace, to the profoundest pit!
> I dare damnation. To this point I stand,
> That both the worlds I give to negligence,
> Let come what comes, only I'll be revenged
> Most thoroughly for my father. (4.5.133–138)

Cool and shrewdly paregoric, Claudius finally calms Laertes to the point where he speaks "Like a good child and a true gentleman" (4.5.150). Then later, playing upon him like a pipe—"Laertes, was your father dear to you? . . . Not that I think you did not love your father" (4.7.107 ff.)—he rouses Laertes' passions once more and directs them toward action:

> What would you undertake,
> To show yourself your father's son in deed
> More than in words? (4.7.124–126)

Claudius knows his man, or his son. The dueling scheme is conceived and carried out despite last second reservations on Laertes' part. Laertes does, in deed, show himself his father's son, as we never doubted him to be. So too in his own roundabout fashion did Fortinbras show himself his father's son, but not before he had first become his own man.

It is no accident, obviously, that one so devoted to the "generall" as Laertes is should ally himself with a man who as "the King" is the generall personified. But then Claudius and Polonius play the same role for Laertes, the King's manipulations merely substituting for those of the father. Indeed, since fatherhood is implicit in kingship, Claudius is a symbolic stand-in for Polonius, ostensibly serving the father's cause while actually securing his own unroyal ends. With the King incorporating the Polonius principle of paternal domination as well as his own secret motives, it is apropos that the dying Laertes should cry "the King, the King's to blame" (5.2.323). Perhaps with these words and with his exchange of forgiveness with Hamlet—which suggests that he is not willing to sacrifice quite everything for his father despite his earlier "I dare damnation"—Laertes gains for himself at last some small measure of personal identity.

FIVE

DEFINITION BY ACTION

HAMLET'S SITUATION is of course more complicated then those of Fortinbras and Laertes. For one thing Laertes and Fortinbras are, through Shakespeare's omission of all maternal references, motherless—a condition, as Hamlet could testify, that goes a long way toward simplifying one's life. This in itself—the one-sided father-son aspect of their stories—alerts us by parallel to the paternal aspect of Hamlet's situation, which we might have scanted had we adopted a Freudian viewpoint or, say, accepted T. S. Eliot's claim that "Mr. Robertson is undoubtedly correct in concluding that the essential emotion of the play is the feeling of a son towards a guilty mother."[11] If we should not overplay the mother-son relationship, as Robertson and Eliot may have done, neither of course should we disregard it. I shall have fairly little to say on that score, however, especially in this chapter, simply because others have said so much.[12] Criticism, one likes to feel, perhaps nostalgically, is a corporate venture, or at least a potluck meal to which each of us brings something, without, we all hope, too much duplication of dishes.

I have been talking about the paradoxical duality of Hamlet's identity as both self and son, the two coalescing in his name, which both is and is not his. For Hamlet to be either "son," through merger with his father, or "self," through division from his father, is made more difficult by his having more than his share of fathers. His first utterance in the play is in response to the word "son." "But now, my cousin Hamlet," Claudius says, adding expansively, "and my son—," at which point Hamlet interjects in aside, "A little more than kin, and less than kind" (1.2.64–65). "Kin" and "kind" (as "species") put the play on universals, yielding the reading, "We are too closely related, having nothing in common." Asked to trade in one father for another, Hamlet is indeed, as he says, "too much in the sun," while his self is too much in the shadow. The only roles available to him in the Council Scene—as either mourner of the dead king or submissive heir of the present king—are filial. The individual self can be expressed only by silence.

If Hamlet at this point is oppressed by "relations" his plight intensifies when the Ghost returns not merely to enlighten him but to impose on him the burden of action. For it is a peculiarity of language that although we can particularize substantives—zeroing in on the *haecceitas*, quiddity, or "inscape" of things or people—the individuality of actions flies untouched through the wide-gauge mesh of our verbal nets.[13] Demonstratives like "this" and "that," and proper nouns of the sort discussed earlier, can be made to single out unique objects—"that man in the corner wearing a yellow tie" or "Beethoven's C Sharp Minor Quartet." But no class of verb corresponds to the demonstrative adjective, and the notion of a "proper verb" ties knots in one's imagination. The most vivid and violent of verbs—say, "slash" or "sneeze"—is doomed by its nature to abstraction.[14] And of course it is precisely that most abstract and universal of all verbs, "to be," that frets Hamlet in his most famous soliloquy—during which the negation of "to be" seems an acceptable escape to him until dreams intrude ("Perchance to dream!"), for dreams issue from particular experience, the subjective, unshareable psyche.

An act, as Macbeth knows even before he himself acts, is defined by its consequences, and a man may be defined by his acts. If Macbeth brings into being the act of regicide, that act reciprocates by bringing Macbeth into being in a new and murderous and, as he himself foreknows, repellent form. When the actor is summed up by his act he inevitably absorbs something of the generic character of action. He who poisons is a poisoner, who lies is a liar, and who revenges is a revenger. Inducted into the category of verbal noun in this fashion (from "proper name," as it were, to "verbal noun" by the agency of action) the actor takes on meaning, especially in the eyes of the law. What he loses in the process is a wealth of "being," the vast inventory of personal features discrete from the act that defines the actor for what he uniquely is.[15] It is this being, or selfhood, that Hamlet is required to sacrifice in order to become a "revenger." Moreover, Hamlet's act is to be performed not for himself, like Macbeth's, but for his father, so that he is doubly subordinated, to the universality of action and to the role of father surrogate.

An act is defined by its consequences. Thus in the "To be or not to be" soliloquy Hamlet ponders the consequences of the ultimate act of suicide, which will be defined by an unknowable "something after death" that "puzzles the will" and thus makes his enterprise "lose the name of action" (3.1.78 ff.). But an act is also defined by its context or scene, as in the familiar legal distinction between the significance of crying "Fire!" in a desert or in a crowded theater. Thus what also puzzles Hamlet's will about the act of revenge is its context, which in the broadest sense is the "world" of Denmark so brilliantly described by Maynard Mack.[16] That context is especially significant in *Hamlet* because unlike Shakespeare's other major tragic heroes, each of whom introduces evil into a world of comparative innocence, Hamlet inherits a world already contaminated by the misdeeds of Claudius and his mother. To come to terms with evil, the heroes of *King Lear, Othello,* and *Macbeth* have largely to come to terms with themselves; an order of recovery must begin with self-knowledge. But Hamlet has not merely himself to come to terms with but also an outside world warped through no act of his—a world miasmal with mystery, disease, degeneration, death, betrayal, and false seeming. To live amid such contamination, merely to suffer a world given over to spiritual rot, is task enough. To have to act, however, and to bear the identity created by one's action, is harder yet. But the Ghost asks even more of Hamlet—that he touch pitch and remain undefiled:

> But howsoever thou pursue this act
> Taint not thy mind; . . .[17] (1.5.85–86)

If men are defined by their acts, and acts by their contexts, then the Ghost's caveat seems a riddling impossibility. Hamlet's predicament is illuminated by contrast with Laertes and Claudius, who in matters murderous care nothing for context. Laertes would prove his father's son by cutting Hamlet's throat "in the church," letting the holy scene define his act and himself however foully; and Claudius nods his agreement, "No place, indeed, should murder sanctuarize" (4.7.126–127).

Hamlet cannot be so unmindful of his scene, or of his precedents.

If he is to kill a king he must be willing to share his identity with all killers of kings. As he contemplates regicide his fictional precedents are chivalric combat in the field on the one hand and murder in the orchard on the other—the open sword of Old Hamlet or the hidden poison of Claudius. Even the first of these is no longer serviceable. The chivalric style of Old Hamlet and Old Fortinbras has already receded into the feudal, almost mythic past by the time Hamlet must act. Once the poisons of Claudius have seeped into Denmark, coursing like quicksilver through "the natural gates and alleys" of not merely his brother's body but the body politic as well, chivalric contests waged under covenants of honor have become hopelessly anachronistic. Claudius is now the current model for regicide, poison and lies the weapons in vogue, and a diseased and rotten world the inescapable context. Now act! the Ghost demands, and "taint not thy mind."

When corruption goes abroad, the sensitive individual usually goes within, seeking sanctuary from the public plague. Romeo and Juliet follow this course, retiring into Juliet's orchard, denying family ties, and striving to meet as purely individual lovers. However, the fact that the orchard in *Hamlet* is the scene of original corruption implies the unavailability of this solution to Hamlet. Nor are the groves of Academe open to him: even before he learns of his father's murder, his intention of returning to Wittenberg is intercepted by Claudius. If, withdrawing, he cannot be himself in Wittenberg, he is assured by Claudius that, remaining, he can "be as ourself in Denmark" (1.2.222). In the shared identity of the common noun "king-killer," Hamlet risks being just that.

However much Hamlet might prefer to remain uniquely himself, isolated from a Denmark whose times are out of joint, he is compelled by "family" to act, to insert himself destructively into the contaminated world of the court. Under normal circumstances, not to act at all would carry its own guarantee of innocence. Given the Ghost's injunction to revenge, however, inaction becomes as fraught with guilt as action. To remain true to himself, free of defilement, Hamlet must of necessity betray his father. To adopt his father's cause, on the other hand, he must betray himself. It is hardly surprising that this dilemma should give him pause.

SIX

Tmetic Structure

LET ME leave Hamlet at this point—puzzled of will, irresolute, beroguing himself for his failures—and examine the claim of T.S. Eliot that "Hamlet's bafflement . . . is a prolongation of the bafflement of his creator in the face of his artistic problem."[18] I have elided from this quote Eliot's reference to the "objective correlative" because I am not concerned with how well Shakespeare reifies emotions in the play but with how he achieves structural unity in it. For the structure of *Hamlet* is, if not perhaps a baffling problem, surely a troubling one for the playwright. How he solves that problem will get us back to Hamlet's problems with his divided identity.

A major structural aim of *Hamlet,* as it is of any play, is implied by Hamlet's address to the players when he urges them to "suit the action to the word, the word to the action" (3.2.19–20). Of course he refers here to acting, but his words apply equally well to drama itself, which requires a marriage of words and action, poetry and performance, dead script and live actors. That marriage (but more immediately its temporary absence) is most apparent in the form of revenge tragedy, where an initial "word" (the hero's vow of revenge) is fulfilled by a terminal "action" (the revenge itself), with the interval between the two devoted to attempts by the hero to bring about the desired fulfillment. The necessity of this "interval" is the subject of some remarks in 1736 by an anonymous critic (probably Thomas Hanmer), who felt that Shakespeare had not mastered his form:

> To speak truth, our poet, by keeping too close to the groundwork of his plot, has fallen into an absurdity; for there appears no reason at all in nature, why the young Prince did not put the usurper to death as soon as possible, especially as Hamlet is represented as a youth so brave, and so careless of his own life.
>
> The case indeed is this: had Hamlet gone naturally to work, as we could suppose such a Prince to do in parallel circumstances there would have been an end of our play. The poet therefore was obliged to delay his hero's revenge; but then he should have contrived some good reason for it.[19]

As the final clause suggests, the usual strategy in dealing with delay is to obscure its presence by providing "some good reason for it"—by engaging the hero in a complicated plot for revenge, perhaps, or by preoccupying him with defensive maneuvers until he is free to act. In any event the playwright would not want his audience distracted from the dramatic illusion by the kind of thoughts expressed by this critic.

And yet Shakespeare, far from obscuring the element of delay, has gone out of his way to publicize both his hero's failure to act and his inability to account for this failure ("I do not know/ Why yet I live to say 'This thing's to do' " [4.4.43–44]). Hamlet's rebukes of himself in soliloquy, the Ghost's return "to whet thy almost blunted purpose" (3.4.115), the contrast of Hamlet with the energetic Laertes and Fortinbras, all combine to underscore the fact that in this play the interval between "word" and "action" *is* an interval, an inevitable structural gap required by the form of revenge tragedy and filled by our makeshift playwright with apparently irrelevant material.[20]

In retrospect we can see that the form of the play, so stressed by Shakespeare, can be likened to several rhetorical constructions—what Puttenham calls "tmesis," "parenthesis," and "parecbasis." The most compressed of these is the scheme tmesis, in which a compound word is separated by an interjection, as in Hopkins' "brim, in a flash, full."[21] Usually of greater length are the familiar parenthesis, in which a syntactic development is delayed by bracketed interpositions, and at the level of argument "parecbasis" or digression, a trope of invention in which, as Thomas Wilson says, "we swarve from the matter, upon just consideration, making the same to serve for our purpose, as well as if we had kept the matter still."[22]

In each of these constructions, as in *Hamlet* itself, the accentuated interval between beginning and end threatens the unity of the expression. As a result, great stress is placed on the end-term, which must rescue the stalled or diverted middle from irrelevance. Of course the middle may not always be rescuable. Thomas Wilson, speaking of digression, says he knew "a Preacher that was a whole hower out of his matter, and at length remembering himself, said well, now to the purpose . . . whereat many laughed, and some for starke wearinesse were faine to goe away."[23] In the preacher's sermon the end evidently issued logically, if not experientially, directly from the beginning, the middle wandering off into its own cul-de-sac. Ideally in this kind of form the end should unite with both the beginning and the middle. Thus in Hopkins' "brim, in a flash, full" the end-term "full" completes the beginning semantically and issues from the middle alliteratively.

And what, then, of *Hamlet?* Its structure requires an act of completion, a final murder that will fulfill the initial vow of revenge. At the same time, if the middle is to be relevant, the act of revenge should be a product of the hero's efforts subsequent to his vow. That, however, is not the case. The killing of Claudius fulfills the initial vow well enough, but it is not a product of Hamlet's efforts during the middle of the play. Indeed, Claudius' death, as Dr. Johnson complains, "is at last effected by an incident which Hamlet has no part in producing."[24] Not that Hamlet has been inactive. He has assumed an antic disposition, mocked Polonius, traduced Ophelia, toyed with Rosencrantz and Guildenstern, partly written and directed "The Murder of Gonzago," killed Polonius, reviled his mother, boarded a pirate ship, and contemplated skulls in the graveyard. But none of these acts has swept his way to revenge, and in fact when he has Claudius at his mercy in the Prayer Scene he spares him, albeit for unmerciful reasons. Finally, after all this digressive action and inaction, we arrive at the Duel Scene, where Shakespeare, rather like Wilson's wandering preacher, apparently remembers himself and says "Well, now to the purpose."

If the play requires an act of completion, the act itself requires an actor or agent. The logical agent, the Ghost, is evidently lacking in the necessary corporeality, and so must employ Hamlet as its instrument. But Hamlet, although willing enough at first, is soon diverted from his goal, in part, as Maynard Mack has argued, by the distractions of his world. Faced with a reasonably well-defined, concrete task—to kill a king—Hamlet expands upon it infinitely as he grows obsessed with the sordid nature of the world he inherits from Claudius and his mother. Looking beyond his particular task, he addresses himself to universal issues, to questions of time and action, of death, corruption, and the various blights man was born for. The point is writ small in the "To be or not to be" speech as the specific question "Should Hamlet commit suicide?" quickly graduates into the universal question "Why should man live in such a world?"

However, it is typical of the slipperiness of universals and particulars that in universalizing his problem Hamlet particularizes himself. For it is during this middle period of the play that Hamlet reveals himself to us as one of the most distinct and memorable characters in literature. In a complex fashion, one might say, Hamlet during this period is *naming* himself. To do so, he must sacrifice his father's cause and pay the price in guilt. That is, "naming" himself involves "de-naming" himself as his father's son and namesake. Like Stephen Dedalus, who defines himself less by merger than by division—by disengaging himself from church, state, family, and school-fellows—Hamlet acts on Spinoza's principle that "All determination is by negation." He asserts not what he is but what he is not, disjoining himself through word and deed from Old Hamlet, Polonius, Ophelia, Laertes, Gertrude, and most of all Claudius. His identity grows complex not by multiplying relationships but by multiplying disrelationships. His extraordinary multifariousness of being would be radically abbreviated if his thoughts and feelings were channeled exclusively into the role of revenger, if "self" were entirely subsumed under "son." And yet precisely because it is not, because the roaming and irresolute self takes priority over the revenge-minded son, Shakespeare's play seems in danger of deteriorating from dramatic action into personal biography, into a study of the Prince without the Play, or at least at the expense of the play. If the Hamlet of the middle is irrelevant and even apparently an impediment to Hamlet the generic revenger, is not the middle of the play itself irrelevant to the overall dramatic structure? Is the middle rescuable?

SEVEN

Naming Hamlet and *Hamlet*: Concrete Universal

TO TALK further about dramatic unity in *Hamlet*, let me adopt some notions from John Crowe Ransom's theory of the "concrete universal." Ransom's position—derived from Kant, Hegel, Coleridge, and perhaps Clive Bell—is that in poems a universal idea or argument, the "structure," is given concrete embodiment in a "texture" of local details. As Wimsatt and Brooks explain:

> Though the texture is strictly irrelevant to the logic of the poem, yet it does after all affect the shape of the poem; it does so by *impeding* the argument. The very irrelevance of the texture is thus important. Because of its presence we get, not a streamlined argument, but an argument that has been complicated through having been hindered, and diverted, and having thus had its very success threatened. In the end we have our logic, but only after a lively reminder of the aspects of reality with which logic cannot cope.[25]

Ransom formulated his notions to account for poems, especially short lyric poems. But if we transpose his theme into a dramatic key the poetic "argument" becomes the plot, the pattern of evolving action, which is "universal" in that it is shared by plays comprising, in the case of *Hamlet*, the revenge tragedy genre. That plot—featuring a crime, a vow to revenge, obstacles to revenge, and a final fulfillment of the vow by a murderous act—is on this view "impeded" by an irrelevant texture. In *Hamlet* this impediment appears in the middle of the play where Hamlet the unique individual receives expression at the expense of Hamlet the revenging son.

Since delay is a built-in feature of revenge tragedy, plays of this sort normally set obstacles before the hero. They do not normally make the hero himself the major obstacle, and they do not have him repeatedly announce that nothing impedes his revenge but himself. Hamlet is baffled by his inability to act, T.S. Eliot says, because Shakespeare is baffled by his own inability to discover an objective correlative adequate to the emotion he needs to express. But perhaps Shakespeare is not so much baffled as he is resistant, and what he needs to express is not so much deeply subjective as conventionally dramatic. Hamlet's expressions of bafflement serve, as I have said, to publicize the structural hiatus in Shakespeare's play and—by announcing that he has cause, will, strength, and means to do it, and yet cannot—to suggest that his delay is at least partly arbitrary, occasioned less by Hamlet himself than by the dramatic structure in which he finds himself. For Hamlet stands not only between himself and his revenge on Claudius but also between the conventional form of revenge tragedy and its fulfillment in Shakespeare's play. And of course it is Shakespeare who has stood Hamlet in that interjacent position. In other words, Hamlet's inexplicably stalled revenge may be to some extent a metadramatic reflection of Shakespeare's resistance to the structural syntax of revenge tragedy. Certainly the humorous catalogue of dramatic forms he delivers by means of Polonius—

> The best actors in the world, either for tragedy, comedy, history, pastoral, pastoral-comical, historical-pastoral, tragical-historical, tragical-comical-historical-postoral, scene indivisible, or poem unlimited . . . (2.2.396–400)

—suggests that Shakespeare has a rather light regard for generic prescriptions.

Let me put it, as I think Shakespeare has, in terms of names. During the period of delay Hamlet is naming himself in all his complex individuality. At the same time Shakespeare is, if not exactly naming *himself*, at least naming his play in all *its* complex individuality. And this dramatic naming involves, like Hamlet's, a "de-naming." For there is one more curious parallel concerning names. Fortinbras is not the only analogue to Hamlet in having a name identical to that of his father. The play *Hamlet* has also, or in its own time *had,* such an ambiguous identity, inasmuch as its dramatic "father," the lost play written around 1589 presumably by Thomas Kyd, was also called *Hamlet,* by which name one of its performances was recorded by Philip Henslowe on June 11, 1594. This play, now known as the *Ur-Hamlet,* is thus related to Shakespeare's *Hamlet*—both genetically as a source and generically as a revenge tragedy—precisely as Old Hamlet is related to young Hamlet.

From this point of view Shakespeare's problem is similar to that of his hero, but not in the manner Eliot proposed. Like Hamlet, Shakespeare has committed himself to a generic task, not an act of revenge but, like revenge (which seeks to restore the past by erasing the intervening time), an act of restoration: the writing of a play based on the *Ur-Hamlet,* which has by this time gone out of fashion but which evidently still remains a popular item of theatrical memory. In adopting Kyd's plot and some of his dramatic techniques, however, Shakespeare risks producing a play that shares not merely the name of its theatrical father but also its very form and pressure. As "Hamlet" both is and is not Hamlet's possession, so the title of *Hamlet* both is and is not Shakespeare's.

How does Shakespeare "de-name" the *Ur-Hamlet?* Obviously by writing a different sort of play. Of course we do not know the contents of Kyd's play, but the repetition of the phrase "Hamlet revenge" (by Lodge in *Wit's Miserie,* by Dekker in *Satiromastix,* and in a pamphlet by Samuel Rowlands) suggests that, like *The Spanish Tragedy,* its focus was on the achievement of revenge rather than on the enervating effects of self-exploration. More than that, though. If self-exploration distinguishes Hamlet from other revengers, then form-exploration distinguishes *Hamlet* from other revenge tragedies. Hamlet's self-chastisements, the Ghost's return, the contrasting revengefulness of Fortinbras and Laertes—all the ways in which Shakespeare publicizes the fact of Hamlet's delay—announce in effect to the audience, "What we have here, you see, is the old revenge tragedy form, with its built-in hiatus between the vow to revenge and the act of revenge, a hiatus which is usually obscured but which I pointedly advertise." Such advertisements, however, are not themselves part of the revenge tragedy genre. They are metadramatic, second-level comments incorporated into the play itself. As such, they constitute an individualizing of the genre—perhaps even a subverting of the genre—in the very process of announcing its existence.

Thus in similar self-reflexive ways, both hero and play become individualized, confiscating for themselves the names they share with their respective fathers. The nominal identity and generic kinship of *Hamlet* and the *Ur-Hamlet* are merely a matrix from which the unique, self-conscious particularity of Shakespeare's play develops. In the process of that development, however, Shakespeare has opened the door of delay so wide that he endangers the unity of his play. How he escapes that danger we shall try to discover shortly.

EIGHT

HAMLET AS CHARACTER/ACTOR

IF SHAKESPEARE'S play as I have just said is form-conscious, it is also as many have said theater-conscious. A touring stage company suddenly appears in Denmark, allusions are made to the child actors of Shakespeare's day, Polonius retails a catalogue of dramatic genres, a player auditions for Hamlet, Hamlet discourses on styles of acting, "The Murder of Gonzago" is performed, and theatrical terms like "act" and "play" are endlessly explored.[26] One effect of this is that the illusions of theatrical art and the pandemic "seeming" of the Claudian court interpenetrate to fashion a pervasive mysteriousness in which the reality of illusion and the illusion of reality dissolve and metamorphose like the forms in an engraving by Escher. This mysteriousness invades not only Hamlet's Denmark but Shakespeare's Globe. Uncertain at times whether drama is a metaphor for life or life a metaphor for drama, we find it hard to divide the play of illusions from the illusion of a play. With Ophelia "acting" for Hamlet in the Nunnery Scene; and Hamlet acting for the King and Polonius; and the King and Polonius, and all, acting for us in the Globe, who are observing the King and Polonius observing Hamlet observing Ophelia, we may well ask how fully the fiction of theatricality engages with our experience in the theater. On the one hand the play seems to draw us into its illusions of life in the Danish castle; on the other hand its explicit metadramatism functions as a Brechtian "alienation device" to erase its illusions and make us see it at least momentarily from a curious aesthetic distance as a purely theatrical construct.

This artistic ambiguity, which has something in common with the reversible figures or visual puns so indispensable to students of perception like E.M. Gombrich, is of course present in Hamlet himself. Indeed any dramatic character can become a reversible image by virtue of his divided identity. For on the one hand he is a more or less realistic person whom we see making choices and taking actions that help create the illusion of life for an audience willing to suspend its disbelief for a time. But on the other hand he is not his own person at all but an instrument of the playwright, nothing but a dramatic role. He does not choose and act, as we had been willing to believe; rather he is made by the playwright to choose and act as he does in order to fulfill the needs of the play. Thus a character is both a source of action, expressing his will, and an instrument of action, dancing to the tune of the playwright and his plot. Normally the playwright will want to fashion a character who serves the plot while appearing realistically to act on his own.

This division of identity can appear in any literary character, whether in play, poem, or novel. Plays, however, give rise to a reinforcing division of identity because in them we can also distinguish between character and actor, between the illusion of a man named Hamlet and, say, John Gielgud, who conveys that illusion to us. If we are absorbed in the play as fictional life, the actor Gielgud is artfully erased, consumed by the character Hamlet. If we become conscious of the actor for some reason (forgotten lines, extraordinary performance, a fit of coughing), then it is Hamlet who is erased, along with the fiction of life in Elsinore, and we are left with an unclothed emperor of theater—greasepaint, boy actors playing female roles, stage props, all the makeshifts of the stage for which Shakespeare repeatedly apologized in the prologues of *Henry V*.

Hamlet the character acts, as it were, on his own volition within the Denmark that constitutes his reality. Gielgud the actor, however, is constrained by plot and script, by the role he is handed. If the actor finds his role unsatisfactory, he may complain to the playwright: "Hamlet would not say that, would he?" or "Why not kill Claudius while he's praying?" And as the actor gets further into his role the "he" gradually becomes "I": "How should I look at Gertrude here?" or "Why, since I have cause and means and strength to do it, must I keep on stalling?"

In Hamlet, Shakespeare seems to have created a character who is, though it seems odd to say so, conscious of his dual identity and able to express both sides of himself, almost as though he were an actor at a rehearsal. He puzzles over the fact that as a character he is fully equipped for revenge but that as an actor, or instrument of the plot, he is not allowed to proceed with it. Thus, envying the player's facility with tears, distraction, and a broken voice, Hamlet says:

> What would he do,
> Had he the motive and the cue for passion
> That I have? (2.2.560–562)

Such a question maneuvers the audience into seeing Hamlet not as the Prince of Denmark but as an actor in the Globe—an actor playing Hamlet, who is envying an actor playing Aeneas. Lamenting that his own role is full of passionate potentials to which he is denied expressive access, Hamlet seems to rebel against his role and, as if to spite his author, clarions passion in all directions, ending in a great spate of invective: "Bloody, bawdy villain!/ Remorseless, treacherous lecherous, kindless villain!" Then, as though embarrassed by this out-of-character, or out-of-role, behavior, he tries to talk his way back into his part:

> Why, what an ass am I! This is most brave,
> That I, the son of a dear father murdered,
> Prompted to my revenge by heaven and hell,
> Must, like a whore, unpack my heart with words,
> And fall a-cursing, like a very drab,
> A scullion! Fie upon't! Foh! About, my brains! (2.2.583–588)

Throughout the middle of the play Hamlet the character is allowed (is of course obliged) to explore the plenitude of his own being, to spin out on the filaments of speech the magic and marvelous web of his personality. The actor-Hamlet is permitted his complaints, his bafflement, his minor revolts, but for the most part he is suppressed. However, if the play is to conclude, Hamlet the singer of self must eventually be silenced; otherwise he could go on endlessly examining himself and his world until an eddying in the dramatic current became a whirlpool. But this transition must be carefully negotiated, because if Shakespeare simply transforms the truant individual into the generic revenger the middle of the play will have been lost to episodic irrelevance. What Shakespeare needs for his digressive structure—to make his "swarving from the matter" serve his purpose "as well as if [he] had kept the matter still"—is an end-term that will both rescue the middle and fulfill the beginning. In the Graveyard and Duel Scenes of Act 5 he provides such an end-term.

NINE

"This Is I, Hamlet, the Dane!"

MOST CRITICS agree that after the ocean voyage we encounter a changed hero, one whose restless questioning of himself and his world, whose inability to remain "content with half-knowledge," has given way to something like resignation, perhaps to Keats's Negative Capability, the ability to be "in uncertainties, mysteries, doubts, without any irritable reaching after fact and reason."[27] The Hamlet who agonized in irresolution between "To be" and "Not to be" has now become the Hamlet who concludes his "readiness is all" speech with the quiescent "Let be." Surely a change of some sort has been dramatically prepared for. Hamlet has been offstage for a little over three scenes. He is dressed differently from when we last saw him, probably in his sea-gown and traveling gear instead of in the disarray of his "antic disposition."[28] He has put soliloquy behind him. He can now admire unpremeditated action, talk of a divinity that shapes our ends, see a special providence in the fall of a sparrow, and address himself to his own death composedly. The question of course is "Why?" It is a question not easily answered, however, for which reason it is not often asked. Hamlet accepts his world, but Shakespeare, as Maynard Mack observes, "does not outline for us the process of acceptance."[29]

Perhaps part of the difficulty lies in the "reversible image" aspect of Hamlet's identity. As a realistic person Hamlet passes from irresolution to "readiness," from self-exploration and cosmic doubt to an acceptance of self and world. Such a change, we feel, should be made psychologically plausible; but because the change takes place offstage, Hamlet's thoughts and feelings are undramatized and somewhat fugitive. In his other mode of being, as the instrument of plot and playwright, Hamlet must exchange the role of truant for that of active revenger. Since roles do not have psyches, the only plausibility required in this connection is formal: Hamlet must be maneuvered to serve the causal demands of the plot and the aesthetic context of the play. In both modes he advances from the particular toward the universal, passing (as person) from self toward son, from individual toward society, and (as role) from structural irrelevance toward contextual integration. As person he now becomes answerable to his ghostly father. As role he becomes answerable to the playwright, his theatrical father (doubly so if Rowe was right about Shakespeare's having acted the part of the Ghost).

Probably no attempt to account for Hamlet's change on a purely psychological basis will prove entirely persuasive. Nor, on the other hand, are we likely to be satisfied by a claim that Shakespeare, having particularized Hamlet during the middle of the play, now simply shifts dramaturgical gears and transforms the truant individual into the generic revenger demanded by the plot.[30] It would seem that Shakespeare has done neither the one nor the other, but rather both simultaneously, in an attempt to align the two dimensions of Hamlet that were previously askew. From the concrete and the universal Hamlets must come the concrete-universal Hamlet. The individual self must unite with the generic son, the person with the role, the character with the actor. Thus the change is effected on grounds partly psychological, partly symbolic.

To begin with, one thing at least seems clear in retrospect—the major moment of the transition, which is arrested for us in Hamlet's account of his ocean voyage at the start of Act 5, Scene 2, when he tells Horatio,

> Sir, in my heart there was a kind of fighting
> That would not let me sleep. Methought I lay
> Worse than the mutines in the bilboes . . .

Pausing here, we might note that these lines imagistically sum up for us the Hamlet we have known prior to the ocean voyage—a Hamlet whose restlessness of spirit precluded resignation; who could not sleep contentedly among the calamities of life or risk what dreams might come in the sleep of death; to whom the world was a prison "in which there are many confines, wards, and dungeons, Denmark being one of the worst" (2.2.246–248); and who has himself in his covert fashion been a "mutine" against Denmark's new command. Out of this irresolution of half-sleep Hamlet suddenly, rashly—"and praised be rashness for it"—rises, fingers the packet kept by his school-fellows, opens it, and discovers the King's lethal commission to England, which he then rewrites and seals, along with the fates of Rosencrantz and Guildenstern.

In retrospect as he tells Horatio the story, Hamlet marvels at the ease and effectiveness of these unpremeditated acts. Rashly he arose, and rashness is to be praised. Rashly he rewrote the commission ("Or I could make a prologue to my brains" [5.2.30]), and his rashness saves his life. On the one hand he acts on impulse to protect himself and to retaliate; on the other hand he feels himself acted upon as part of a larger order of causation. Within the realistic illusion of life in Denmark, Hamlet the person senses an overriding "divinity that shapes our ends,/ Rough-hew them how we will" (5.2.10–11). Within the theatrical dimension of the play, as suggested by the lines "Or I could make a prologue to my brains,/ They had begun the play," Hamlet the actor is being called upon to fulfill his role in the plot. In both dimensions, Hamlet the individual is beginning to take a subordinate place within a larger context—the providential plot that governs human experience in Denmark and the revenge tragedy plot that governs dramatic experience in the Globe theater.

The focal source of Hamlet's change, the major prop of his shipboard drama, is the King's commission to England. If the earlier Hamlet particularized himself to some extent by pondering universal issues, the universal now becomes most immediately particular when he reads the commission and looks upon his death warrant. Death no longer appears abstractly as an unknowable Not To Be but is concretely imagined as the crush of flesh and bone by the stroke of an unsharpened axe as the head—*his* head—rolls and spurts under the studious gaze of his two school-fellows. Yet this personalizing of death, as though the reading of the commission has transformed "death" from an abstract noun into a proper name, seems less awful to Hamlet than the universalized death of his famous soliloquy. It is as though Hamlet has already come to terms with death before the negotiations have begun, or as if the effect of the sea voyage experience (i.e., his reconciliation to death) somehow governs the experience itself.

If cause and effect seem reversed here, we can attribute it to Shakespeare's own dramatic reversals and interventions, which are of two kinds. First, the moment of Hamlet's crucial change is not dramatized for us but narrated. The change itself takes place in that great off-stage void from which only ghosts, ambassadors, and pirate messengers return. As narrated history, then, Hamlet's nascent change appears in retrospect, filtered to us through time the distancer and Hamlet the tale-teller who is himself at one lucky remove from his experience. Thus the changing Hamlet at sea lives only in the voice of the changed Hamlet back at court, and the tone of that voice imparts a sense of composure and transcendence (literally providential transcendence) to an action described as "rash" and frightening.

It is understandable that Shakespeare would choose to narrate rather than to dramatize the sea-voyage incident, if only to preserve the "unity of place" in a play that deals so cavalierly with the unities of time and action. His second intervention, however, is more curious, since it involves an inversion of the dramatic syntax of Act 5. Why, we must wonder, does Shakespeare let Hamlet converse with Horatio throughout the lengthy Graveyard Scene (5.1) without once mentioning, as we should have expected him to do immediately, the extraordinary events of his sea-voyage? Why, that is, is the story of the King's commission not told until the beginning of Act 5, Scene 2?[31]

Perhaps it is for the same reason that Shakespeare narrates rather than dramatizes the sea-voyage adventure, for by inverting his dramatic syntax in this manner he causes the *story* of the sea-voyage to issue from the Graveyard Scene as much as from the adventure itself. As a result, Hamlet's experience of "death" aboard ship, or rather his telling of it, is colored by his experiences in the graveyard, where we see him registering, in a kind of thanatophany, the corporeal reality of death in both its universal and its particular aspect. The process is summed up in that moment, the most emblematic of the play, in which Hamlet contemplates the skull of Yorick. The skull itself is one of two tossed up earlier by the singing Gravemaker as he fashions his long house for Ophelia. Hamlet has conceived of these skulls in the most universal terms, imagining them to have belonged to a politician, a lawyer, a lord, even to Cain. However, when the Gravemaker says "This same skull, sir, was Yorick's skull, the King's jester," Hamlet replies with an astonished "This?" and takes the skull in his hand (5.1.198–200). The graduation of the skull from an undifferentiated universal to a concrete particular is mirrored verbally by the concrete universality of Hamlet's "This?" For the demonstrative "this" is at once a class-of-all-classes term, capable of referring to anything at all, and a precise particularizer singling out one unique object. In isolation, "this" encompasses everything and distinguishes nothing. But in context—at this time, in this place, uttered by Hamlet, the skull in his hand—"this" has the verbal focusing powers of a microscope. Then, from the "particularizing half" of Hamlet's "This" there flows, in macrocosmic form, his famous speech, "Alas, poor Yorick!" in which Hamlet's imagination, like My Lady Worm in reverse, fleshes out the anonymous bare bones once again with lips to kiss and a back to bear small princes on and flashes of merriment to set the table on a roar.

This particularizing of death in the Graveyard Scene—the process with Yorick's skull is repeated in larger compass when the Gravemaker's literally precise but unrevealing "One that was a woman, sir; but, rest her soul, she's dead" is gradually transformed in Hamlet's stunned consciousness to "What, the fair Ophelia!"—thus becomes by virtue of Shakespeare's syntactic inversion the dramatic "cause" of Hamlet's narrative conversion of "death" from an abstract noun to a proper name in the following scene, though of course in actual chronology the latter preceded the former.

Hamlet's shipboard encounter with death is a culmination of that whole middle section of the play in which he individualized or named himself. The final step in his naming the "I" in "Hamlet" is his naming of death imaginatively as "mine." It marks his completion as an individual. But it also marks a transition from the individual to the generic avenger. For the Hamlet who makes death his own is saved from that death by assuming his father's authority, as he indicates when Horatio asks him how the rewritten commission was sealed:

> Why, even in that was heaven ordinant.
> I had my father's signet in my purse,
> Which was the model of that Danish seal;
> Folded the writ up in the form of the other,
> Subscribed it, gave't the impression, placed it safely,
> The changeling never known. (5.2.48–53)

The fact that Hamlet takes decisive action in his own cause, that he employs his father's signet to seal that action, and that he assumes royal authority in the process suggests that Hamlet the self, Hamlet the son, and Hamlet the royal Dane all cooperate in the saving of his life. This multiple identity will be confirmed by Hamlet very shortly. At this point we see that both of Shakespeare's interventions have conspired to particularize Hamlet's experience at sea and yet to subsume that particular under the aegis of the universal.

In the graveyard we encounter a Hamlet who is still changing. The graveyard is itself a scene of changes, reminding us of the parts we all play in the evolving plot of our lives; and the first words we hear, from the gravemaking clowns—a prelude to Hamlet's entrance—sound the theme of agents and instruments. The question is whether the mad Ophelia drowned herself willfully or accidentally. The answer of the churlish Priest is that she drowned herself willfully and should therefore, as agent of her own destruction, lie in unsanctified ground, sans prayer and requiem. However, "great command o'ersways the order" of the Priest, enabling Ophelia to receive, if not full absolution, at least the ambiguity of "maimed rites."

If, as the Priest claims, Ophelia's "death was doubtful," leaving us unsure whether she was willful actor or passive agent, Hamlet increasingly regards all life and death as instrumental episodes in a larger plot. Those whose skulls he examines in imagination—the politician, the courtier, Lord Such-a-one, the lawyer, and even Yorick—play their parts in their individual dramas and make a common exit to the tiring house of soil. Even death is but a phase in the continuing process. In a still more comprehensive plot Alexander exits from life only to enter upon another series of roles, as noble dust, earth, loam, and at the vulgar end as beer-barrel stopper.

Hamlet senses that he too has become part of a larger process: the plot of Providence as scripted by the divine Playwright. The events of his life, he now seems to feel, have become incorporated into universal history. This would appear to be inconsistent with my earlier claims. That is, I suggested that the radical particular, the undefined "This that simply is," acquires meaning by becoming a member of a class ("This . . . is a sparrow") but that in doing so it loses "being," the plethora of individual characteristics that are disregarded in the abstraction of essential class features. This loss of being or selfhood increases with the degree of abstraction—as "this" graduates from "sparrow" to "bird" to "vertebrate" to "animal" to "organism," etc.—which is what we saw Hamlet resisting during the middle of the play as he became instead a character of intrinsic biographical interest, unfamilied, an end in himself. Yet now, in an apparent reversal, he regards himself as a member of the most universal of classes, one of God's creatures, subject to His order. Hamlet's uniqueness seems about to blur into the great generality of things.

However, Hamlet's references to transcendent order in Act 5 suggest not so much a surrender of the concrete to the universal as a fusion of the two. For in the divine dispensation, not the smallest action—Hamlet's rash theft in the night, his use of his father's signet, not even the fall of a sparrow—loses its special character, so particular is God's universal concern. Indeed God seems to specialize in detail work. He reserves a role for the sparrow in His providential drama, and He fine-chisels the rough-hewn ends of men's affairs. In such a scheme the reality of the individual is by no means forfeit to abstraction.[32]

Thus Hamlet accepts his role in the providential play uncomplainingly, in keeping with Sir Walter Raleigh's admonition a few years later in the Preface to his *History of the World* (1614):

> For seeing God, who is the Author of all our Tragedies, hath written out for us, and appointed us all the Parts we are to Play, and hath not, in their distribution, been partial to the most mighty Princes of the World . . . why should other Men, who are but as the least Worms, complain of wrongs?[33]

Hamlet accepts his role as one of the Princes of the World and at the same time his role as one of the Players in the Globe. Earlier, to pursue Raleigh's theatrical metaphor, Hamlet had aspired to the status of dramatist, seeking to attain the kind of synoptic overview that would enable him to comprehend the entire play instead of contenting himself with the limited perspective vouchsafed him as a character. Indeed, for a moment he became a playwright, converting "The Murder of Gonzago" into "The Mousetrap" and thus controlling the destinies of not only the characters within that play but Claudius as well.[34] With his return to Elsinore after the sea-voyage, however, Hamlet begins to acquiesce to the dramatic context rather than trying to dominate and transcend it.

Shakespeare signifies this transition once again with a name. When Hamlet issues his challenge to Laertes (and Claudius) at the grave of Ophelia—"This is I/ Hamlet, the Dane!" (5.1.257–258)—his announcement is a model of self-definition. The anonymous "This" becomes the unique "I," who is part of the paternally shared "Hamlet," who merges with the universal "Dane." During the middle of the play Hamlet could have gone no further than to say "This is I." Now, as he is about to descend with Laertes into the grave and invite death, the presence of all three terms implies the accommodation of the confirmed self within a universal context.

TEN

Definition by Action: Double Revenge

HAMLET'S "This is I,/ Hamlet, the Dane!" marks the end of the hero's naming of himself, his verbal self-definition. He now seems ready to confirm that identity by means of action. But perhaps naming and acting are not as discrete as that sounds. If a man can be defined by his actions, is not the reciprocal also true, that to define oneself is to predispose oneself to action? Naming from this standpoint is not merely symbolic action, a substitute for doing, but incipient action, a preparation for doing. If so, then Hamlet's naming of himself at the grave of Ophelia should contain *in potentia* the act of revenge. That this is so is made more evident when Hamlet, having told Horatio of the ocean voyage, goes on to say,

> Does it not, think thee, stand me now upon—
> He that hath killed my king and whored my mother,
> Popped in between the election and my hopes,
> Thrown out his angle for my proper life.
> And with such cozenage—is't not perfect conscience,
> To quit him with this arm? (5.2.63–68)

Hamlet's revenge will be justified by Claudius' evil acts, which correspond precisely to the three terms of self-definition in Hamlet's "This is I,/ Hamlet, the Dane!" The primal act of killing "my king," which paved the way to the whoring of "my mother," brings into focus Hamlet the son who bears his father's name. Claudius' stealing of the election for kingship emphasizes Hamlet as Prince of Denmark, "the Dane." And the angling for his "proper life" calls forth Hamlet the unique individual. Thus the revenge-minded hero is warranted to act as "I," as "Hamlet," and as "Dane"—self, son, and prince.

Having been rendered potential through naming, the revengeful act must now be made actual through performance. Shakespeare stages that performance most carefully. Despite Hamlet's praise of "rashness" he does not, like Laertes, simply storm the King's quarters with bloody intent. Nor on the other hand does he devise Machiavellian plots and stratagems to effect his revenge. To do so would be to accept Claudius' early invitation, "Be as ourself in Denmark" (1.2.122) and to become most deeply tainted by the world of seeming that has repelled him throughout the play. (Even Claudius admits to Laertes that their own plot is likely to succeed because Hamlet, being "most generous, and free from all contriving," will not examine the foils [4.7.135].)[35] To the contrary, Hamlet has forsworn plots, having come to feel that "our indiscretion sometimes serves us well/ When our dear plots do pall" (5.2.8–9). Instead of a plotter, Hamlet is now a player, with parts in other plots—that of Providence, that of Shakespeare, and, unwittingly, that of Claudius and Laertes. Much earlier, when he ended his speech to the players before the performance of "The Murder of Gonzago," he told them "Go, make you ready" (3.2.45). Now he is himself a player and can acknowledge that in life as well as in theater "the readiness is all" as he awaits his personal cue (5.2.220). That cue comes when he is called upon to play his part in Claudius' plot, a role that requires him to "fall to play" with Laertes.

Immediately before this final "play," however, Hamlet makes a gesture of apology to Laertes:

> What I have done
> That might your nature, honour, and exception
> Roughly awake, I here proclaim was madness.
> Was't Hamlet wronged Laertes? Never Hamlet.
> If Hamlet from himself be ta'en away,
> Then Hamlet does it not, Hamlet denies it.
> Who does it, then? His madness. If't be so,
> Hamlet is of the faction that is wronged;
> His madness is poor Hamlet's enemy. (5.2.228–237)

This is an odd apology in part because the offence is left unidentified as simply "it" and "What I have done." Perhaps Hamlet refers to his outburst at the grave of Ophelia, which Claudius and Gertrude both labeled "madness" (5.1.272, 284). But Hamlet's apology seems excessive to that somewhat peripheral occasion, and I think Shakespeare—by keeping the offence vague, by having Laertes speak of his motives for "revenge" (5.2.242–244)—intends us to see through the business in the graveyard to the primary offence, Hamlet's murder of Polonius. For it is certainly that, rather than the scuffling in the grave, for which Laertes is already plotting revenge.

Hamlet assigns all blame to his madness, not because he is still pretending an "antic disposition" (he says that he *was* mad, not that he *is*)[36] and not because he is making a literal claim to madness ("madness" is less a lie here than a metaphor). The apology repeats the theme of identity—"Hamlet" and "not-Hamlet," Hamlet sane and Hamlet mad—that we have marked in the division between self and son. In these terms Hamlet's madness issues from his identity as "son." In a metaphoric sense Hamlet has been driven "mad" by his father's Ghost, inasmuch as he has been obliged to become, as he told Gertrude, "mad in craft" to conceal his designs from Claudius, whose villainies were revealed by the Ghost. It is Hamlet the son, not Hamlet the unrelated self, who stabs Polonius, thinking him Claudius ("Is it the King? . . . I took thee for thy better" [3.4.27, 33]), who in the same scene extolls his father and denigrates his successor, and who uses words like daggers to pierce his mother's conscience. The equation of madness with Hamlet the son is made explicit with the reappearance of the Ghost, whose special relationship to Hamlet is underscored by its invisibility to Gertrude. As Hamlet says "Do you not come your tardy son to chide" Gertrude cries "Alas, he's mad!" (3.4.109–110). As a metaphor for Hamlet's bond to his father—for that sense in which Hamlet as revenger is "possessed" by the ghost of his father—Hamlet's madness is truly no part of himself, and is in fact "poor Hamlet's enemy." For Hamlet's personal identity and fullness of self lie in his unrelated particularity, in that concrete multifariousness of "being" which is in excess of and irrelevant to his universal "meaning" as son and revenger.[37]

Just prior to the duel, then, Hamlet seeks to dismiss as madness that part of himself that sought to revenge his father and that killed Laertes' father instead. His apology asks for a forgiveness that would erase that murderous act and at the same time divest both young men of their "sonship" by exorcising the "ghosts" of the dead fathers, whose memories demand revenge. That is, a genuine acceptance of Hamlet's apology would erase the "mad," unreal, revengeful "son" who mistakenly killed Polonius and at the same time the other "son" who is honor bound to avenge that death, thus enabling Hamlet and Laertes to meet as individuals unencumbered by family. For a brief moment, then, as the play tips toward its fatal conclusion Shakespeare recalls for us the Hamlet of the middle of the play, the self-searching "I" to whom has subsequently been added "Hamlet, the Dane." In the process Shakespeare draws our attention to Hamlet's ambiguity as both active agent and passive instrument. For the effect of his apology is to transform the autonomous Hamlet, who is defined by and responsible for his own actions, into the instrumental Hamlet, who is possessed by madness and victimized by actions he is compelled to perform. This ambiguity is to be sustained in the duel and brought to culmination in the murder of the King.

If Hamlet the revengeful son is "not himself," neither is Hamlet the duelist, for much the same actional reason. Since in dueling the ordinary free play of movement and bearing, of how a man acts, is disciplined to the conventions of swordsmanship, the man turned duelist loses his individuality in his actional role, as in a more complex manner the actor loses his individuality in his dramatic role. In the duel-drama metaphor which many critics have noted, Hamlet enters a fictional world—an "as-if" world of bated foils in which fencers pretend to engage in real combat—and there loses himself in a series of stylized actions. But of course Laertes is not, in this sense, "acting" (although in another sense he is, since he is disguising murderous intentions behind the role of a fencer pretending to have murderous intentions). Nor, once wounded, is Hamlet. In the form of the unbated foil, reality invades fiction. The illusion of play is shattered, fencing becomes fighting, and at the end Hamlet leaps from his role as duelist into his role as revenger, out of fiction into reality—like an actor leaping from the stage into the box seats—to kill the most attentive member of his audience.

At last the deed we have so long awaited is performed and the promise fulfilled. And yet, we are obliged to ask, is this really the deed we have awaited? Why, after all, does Hamlet kill Claudius? To ask such questions at this late date seems somewhat perverse, given all the motives for revenge Hamlet has collected throughout the play and so recently inventoried. But the reason for asking it is that Hamlet does not in fact kill Claudius as an act of revenge for his father. When he cries "The point envenomed too!/ Then, venom, to thy work" and thrusts at Claudius, it is not poison in the ear but venom on the swordtip that prompts his attack. Hamlet kills Claudius, in short, not for his father but for himself, not in response to the Ghost's command but in direct retaliation for the attack on his own life. Indeed, his stabbing of Claudius is less an act of calculated revenge than a reflex action. Hamlet answers the thrust of Claudius' plot against him as he answered the thrust of Laertes' unbated foil, with a quick and fatal riposte.

There remains another question to be asked: Why does Hamlet kill Claudius twice? For having already stabbed the King with a foil whose venomed point, he has just been told, yields "not half an hour of life," Hamlet then seizes the poisoned stoup of wine and—

> Here, thou incestuous, murderous, damned Dane,
> Drink off this potion! Is thy union here?
> Follow my mother! (5.2.327–329)

Whereas the venom on the unbated foil has been applied by Laertes (4.7.141–149) the poison in the stoup of wine is contributed by Claudius. "He is justly served," the dying Laertes declares, "It is a poison tempered by himself" (5.2.329–330)—a statement that summarizes the long process of self-poisoning in Claudius that began with that first fatal moment in the orchard. Thus the stabbing of the King, though it involves poison, does not issue from the original poisoning of Hamlet's father but from the poisoning of Hamlet himself in the duel. With this first killing of the King, Hamlet fulfills the initial part of his cry of self-definition, "This is I!"

With the second killing he fulfills the remainder of that cry, "Hamlet, the Dane!" The second killing, given the deadly sufficiency of the first, is less a functional necessity than a symbolic formality. It addresses itself to the familial and political aspects of Claudius as the "incestuous, murderous, damned Dane." Hamlet stabs Claudius for himself, but poisons him for his father—and for his mother, whom Claudius is to "follow" now for having followed earlier. The "union" is the pearl Claudius dropped in the stoup of wine, but in its punning sense (as "marriage") it harks back to the incestuous union brought about by the poisoning of Hamlet's father. The suggestion of a "poisoned marriage" pun in the juncture of "poison" and "union" here would be neatly borne out in full if the pearl were set in a ring containing poison in the recommended Lucrezia Borgia style.

ELEVEN

Tmetic Unions

IF WE back away from the scene of death a bit we can see that by having Hamlet kill the King twice, by introducing a strictly unnecessary second killing, Shakespeare foregrounds his having accomplished two other "unions"—the union of Hamlet's previously divided identity and the union of the play's disjunctive dramatic structure. In the tmetic form of *Hamlet* the first killing of Claudius proceeds not from the beginning of the play, not from Hamlet's vow to avenge his father's death, but from the middle. It issues from that long, apparently digressive phase in which Hamlet, distracted from his generic task of revenge, exhibits for us the unique workings of his tormented consciousness—the period in which he "names" himself. Were we presented only with the second killing, with the act of revenge for a murdered father, the middle of the play should have been irrelevant indeed, and Hamlet's conversion from self to son would have seemed inexplicable and arbitrary. As it is, however, in acting for himself Hamlet acts for a self that has been freighted with value and meaning by virtue of that self-exploratory middle phase. His self-revengeful killing of Claudius emphasizes the achievement of individuality that, as we have seen in the case of Fortinbras, is a precondition of meaningful action in *Hamlet.*[38]

If this first killing helps rescue the distinctively individual form of Shakespeare's play by binding the end to what had seemed a purely narcissistic middle, the second killing accedes to its generic form as revenge tragedy by binding the end to the beginning. It is performed by Hamlet the revenger fulfilling his vow to the Ghost and thereby suiting the initial "word" to a terminal "action" in the prescribed revenge play fashion. And just as Shakespeare had repeatedly advertised the apparent irrelevance of the middle of the play by emphasizing the fact of Hamlet's delay, so now he advertises the structural necessity of the second killing by emphasizing its pragmatic irrelevance.

What has made *Hamlet* innovative as a revenge tragedy, a play uniquely itself within its genre, is its departure from the usual generic pattern in order to explore both the consciousness of its hero and the nature of its own form.[39] By this means both play and hero distinguish themselves from their patronyms, the *Ur-Hamlet* and Old Hamlet, and name themselves as discrete entities. Yet as part of the paradox of the concrete universal, individual fulfillment may coincide with generic achievement: self and son may interpenetrate. For the Hamlet who kills Claudius for himself inevitably kills his father's enemy as well, rather like Fortinbras regaining his father's lands in Denmark by conquering for himself in Poland. Perhaps this gives scope to Polonius' famous advice to sons:

> This above all: to thine own self be true,
> And it must follow, as the night the day,
> Thou canst not then be false to any man. (1.3.78–80)

By exploring and naming the "I" in "Hamlet," then, the hero proves at the end not to have been false either to his father or to himself. Earning the name of "Hamlet" for himself, investing it with his own meaning, he has the more truly merited his father's legacy. The singular name is now a symbol for the fusion of self and son wrought by a final act that is at once individual and generic, concrete and universal.

What applies in this regard to the hero applies also to his play. Having gone its own self-defining route, the play nevertheless honors its genetic and generic obligations at the end. It defers to the form of revenge tragedy, and yet in the process it compels that form to make its way—hindred and diverted, its success in constant doubt—through the densities of concrete experience. So wrought by its particular embodiment in this play, in this hero, the generic form is no longer what it was. It has been twisted, tested, and modified by a playwright who in making the form answer to the needs of his peculiar imaginative vision has revivified it as a genre. Thus Hamlet the hero and *Hamlet* the play both achieve the independence of "being" attached to proper names and at the same time the breadth of "meaning" possessed by common nouns. Both have truly named themselves.

PART II

To Be and Not To Be: The Range of Negation in *Hamlet*

TWELVE

Verbal Presence: Conceptual Absence

IN A classic article on *Hamlet* some years back Maynard Mack first called attention to the fact that Hamlet's world "is preeminently in the interrogative mood. It reverberates with questions, anguished, meditative, alarmed."[1] Many of these are questions of fact or identity—"Who's there? . . . Have you had quiet guard?" (1.1.1., 10) Some are questions of value: "And yet, to me, what is this quintessence of dust?" (2.2.320) And some focus upon meaning, or more often upon the absence of it:

> What may this mean . . . ? (1.4.51)
>
> What means your lordship? (3.1.107)
>
> Will they tell us what this show meant? (3.2.141)
>
> Alas, sweet lady, what imports this song? (4.5.27)
>
> What should this mean? . . .
> Or is it some abuse, or no such thing? (4.7.48–49)

The voices asking within the play for meanings seem to speak for the play itself, calling for interpretation. Very few critics have been able to resist that call; and of those who have hearkened to it fewer still have escaped the feeling expressed by Laertes in reply to Claudius' query about the meaning of Hamlet's letter: "I'm lost in it, my lord" (4.7.55).

As a maker of mazes in which we all lose ourselves Shakespeare exhibits a truly Daedalean cunning. However, if he does not always let us out of his mazes, at least he often advertises where the entrances are and graciously asks us in. His curious thriftiness with names, for instance—his unwillingness to spend more than the name of "Hamlet" on the old king, the young prince, and the play itself—seems to address us with sly innocence, asking "What might this mean, my lords?" Unable to resist so gentle an invitation, in Part I, I set off into the labyrinths of the play, working blithely down corridors of names toward the distant sound of shuffling hoofs.

I offered in Part I a fairly positive, determinate, "centered" interpretation of the play. In Part II I want to suggest some of the ways in which the play resists positive interpretation, especially by its employment of the negative mode. Perhaps the most enticing entrance to this winding matter appears near the end of the Closet Scene when Gertrude, having been fairly thrust through by the verbal daggers of her son, asks Hamlet "What shall I do?" and is told:

> Not this, by no means, that I bid you do:
> Let the bloat king tempt you again to bed,
> Pinch wanton on your cheek, call you his mouse,
> And let him, for a pair of reechy kisses
> Or paddling in your neck with his damned fingers,
> Make you to ravel all this matter out,
> That I essentially am not in madness
> But mad in craft. (3.4.188–195)

Why, we can hardly help wondering, such paradoxical advice? Why the conspicuous double negative and the loathsome imagery when Hamlet might simply have said "Tell the King I am truly mad"? To be sure, Gertrude's moral retardation calls for extreme remedies, as the entire Closet Scene has demonstrated, and it may be that the very oddness of Hamlet's advice helps it take root, since in the following scene when the King asks "How does Hamlet?" Gertrude affirms that he is "Mad as the seas and wind, when both contend/ Which is the mightier" (4.1.7–8). Unfortunately, it is not clear whether Gertrude is deliberately deceiving the King here, as Hamlet asked, or simply reverting to her complacent assumption throughout the Closet Scene that "not [her] trespass, but [Hamlet's] madness [was speaking]" (3.4.146).

But perhaps Hamlet is not so much instructing Gertrude in that speech as engaging in a type of word-magic. Seeking to control an uncertain future as well as an unreliable mother, he forecasts a repugnant scene of betrayal and, by enclosing it within the frame of the double negative, erases it. If this is Hamlet's intent, his task would be made easier were he a devotee not of drama but of a nonverbal art like sculpture. Then he could fashion replicas of Gertrude and Claudius in bed and, like a primitive with effigies of his enemy, destroy them. Or, to come at the matter more directly, which of course is not Hamlet's style, he might simply wait for a moment when effigies and words are both unnecessary, as he contemplated doing in the Prayer Scene:

> When he is drunk asleep, or in his rage,
> Or in the incestuous pleasure of his bed,
> At game a-swearing, or about some act
> That has no relish of salvation in it—
> Then trip him . . . (3.3.89–93)

Claudius and Gertrude in bed are hardly the same as effigies of Claudius and Getrude in bed, but both are alike at least in their destructibility. As part of nonverbal reality, both are positive presences that can be, not negated, but eliminated. Commenting on Bergson's "The Idea of Nothing" in *Creative Evolution*, Kenneth Burke says:

> Bergson points out that there are no "negative" conditions in nature. Every situation is positively what it is. For instance, we may *say*, "The ground is *not* damp." But the corresponding actual conditions in nature are those whereby the ground *is* dry. We may say that something "is not" in such and such a place. But so far as nature is concerned, whatever "is not" here is positively somewhere else; or, if it does not exist, then other things actually occupy all places where it "is not."[2]

Things in nature and replicas of them in the visual arts are condemned either "to be or not to be." So too with things a bit out-of-nature, like the Ghost. Gertrude exemplifies Burke's final statement perfectly when Hamlet, staring at the Ghost, cries to her "Do you see nothing there?" and she says "Nothing at all, yet all that is I see" (3.4.131–132). If the Ghost is not there for Gertrude, then other things occupy the space where it "is not." Lacking the duplicity of the negative, kings, queens, effigies, even ghosts cannot both be *and* not be at the same time. That is a delicate skill possessed only by words.

Negation, that is, introduces a paradox into language: the verbal presence of conceptual absence. Such a paradox probably goes unnoticed in a sentence like Hamlet's "It is not nor it cannot come to good" (1.2.158), because "It" in the first clause and the "coming to goodness" of "it" in the second clause have an imageless pallor to begin with, even before the bleach of the three negatives is added. But in constructions where the negated term is more vivid, the negative itself may fade into abstractness. "If you would form an idea of nothing," Burke observes, "you require an image—and, as Bergson points out, an image must be of something."[3] Even in a simple negative like "The castle does not exist" the logical explosiveness of "not," which should blow the pretentious castle out of existence, is itself nullified in some degree by the fact that the castle is at least vaguely imaginable and the "not" is not.

But the paradox of negation is at its most duplicitous in Hamlet's advice to his mother. Here the double negative functions as a conceptual eraser of all that follows it. This suggests how tenuous and vulnerable language is in the presence of the negative. Any sentence, of whatever mass and charge its rhetorical forces, can be ambushed and slain by a single "not." On the other hand, language cannot survive without the negative, which is, Burke argues, "the essential distinction between the verbal and the nonverbal,"[4] and which insures that names are not things, as Plato said, and that signifiers are not signifieds, as Saussure said.[5] However, if language cannot survive without the negative, words can. Hamlet's repugnant images not only resist erasure but take on a verbal presence so graphic and memorable as almost to erase the double negative itself. We may register the logic of Hamlet's "Not this, by no means" and discount the existence of what he then describes, but we can by no means vitiate the life of his utterance or annul our imaginative experience of it. Inspirited by poetic imagery, Hamlet's words about the "bloat king" are more durable than the king himself. You can, after all, kill kings. Even Hamlet in his own good time can kill a king. You can also kill speech: kill the king and his speech dies with him. What you cannot do, it seems, is kill words, for your only weapon is other words. Killing words with words is like observing electrons through an electron microscope. In their quantum shyness the electrons will not reveal themselves fully (position and velocity at once) even to their own subatomic kind, nor will words yield outright to other words. The result is a linguistic version of Heisenberg's Uncertainty Principle.

Negation, then, could be regarded as a suicidal impulse within a language that, like the Everlasting, has fixed his canon 'gainst self-slaughter. The negative cannot destroy without at the same time creating something to destroy. In doing so, it gives life to what it kills.

Let me return to Hamlet's advice to Gertrude a bit further on. Thinking of negation as a logocidal impulse should sidetrack us for the moment to Hamlet's "To be or not to be" speech, where his own impulse to suicide encounters the negative:

> Who would fardels bear,
> To grunt and sweat under a weary life,
> But that the dread of *something after death*,
> The *undiscovered country* from whose bourn
> No traveller returns, puzzles the will
> And makes us rather bear those ills we have
> Than fly to others *that we know not of?* (3.1.77–83)

Life, whose repellence makes suicide inviting, encounters Not Life, that which "we know not of," which repels Hamlet even more forcibly. It is not the presence of death that Hamlet fears. He is truer to the nature of the negative in this respect than Claudio in *Measure for Measure*, who in his famous speech to Isabella first says "Ay, but to die, and go we know not where" and then graphically shows that he *does* know where:

> To lie in cold obstruction and to rot;
> This sensible warm motion to become
> A kneaded clod, and the delighted spirit
> To bathe in fiery floods, or to reside
> In thrilling region of thick-ribbed ice;
> To be imprisoned in the viewless winds,
> And blown with restless violence round about
> The pendent world; or to be—worse than worst—
> Of those that lawless and incertain thought
> Imagine howling—'tis too horrible! (3.1.118–128)

Claudio's imagination is well stocked with images of death. Hamlet's, on the other hand, is prodigal toward life:

> the whips and scorns of time,
> The oppressor's wrong, the proud man's contumely,
> The pangs of despised love, the law's delay,
> The insolence of office, and the spurns
> That patient merit of the unworthy takes . . .

As in Hamlet's advice to Gertrude, the world of Not-Being, which would destroy the world of Being through the act of suicide, gives it instead a verbal life rank with repugnant images.

Claudio is repelled by vivid images of what death *is*. For him death is not negative but a positive presence, a region of strange and horrible experiences into which the spirits of the dead are cast. Hamlet, on the other hand, is repelled by images of what life is. If he gives death a kind of presence as a "country," it is nevertheless an "undiscovered country," a "something" so undefinable as to be a nothing. What he fears is not so much death as life-in-death:

> Perchance to dream! Ay, there's the rub;
> For in that sleep of death what dreams may come,
> When we have shuffled off this mortal coil,
> Must give us pause. There's the respect
> That makes calamity of so long life.

Hamlet is not afraid that death will erase life but that it will not do so sufficiently—that death's nightmares will be simply an intensification of life's bad dreams. Life, though lost, may be found again in death, even as in his speech the imagery of life retains its vigor beside the negative debility of "that we know not of." There's the respect that makes calamity, and words about calamity, of so long life. For however specious it may be for the philosopher to reify the negative—and the risks he takes in regarding *le neant* in this fashion are exemplified by the fate of Polyphemus at the hands of "Nobody"—for the poet, "Not To Be" inevitably entails "To Be."

THIRTEEN

UNCREATIONS

TO RETURN to our textual crux: what, then, is Hamlet doing in his speech of advice to Gertrude? Many things, no doubt—some of them contradictory, some of them hard to distinguish from what Shakespeare is doing by means of Hamlet's speech. After all, the negative principle of "Not This," which in poetic employ becomes "Not This But Nevertheless This," would appear to license contradictions, ambiguities, multiplications of meaning. One could argue that Hamlet is indulging a voyeuristic Oedipal imagination within the self-exonerating safety zone of the double negative. Or that he is sublimating in repugnant but at least nonmurderous language the matricidal impulses that were frustrated earlier in the scene by the arrival of the Ghost. Or that he is trying through magical force of speech to control the future. Or, to the contrary, that he is negating his own negations. His repellent images, that is, may be intended, not only to make Gertrude recoil into goodness, but also to express his own futility in a world so replete with evil that goodness itself withdraws into the unexpressive vacancy of "Not this, by no means."

Pursuing this latter line, we might regard Hamlet's speech as a reaction (by Hamlet, Shakespeare, or both) to Gertrude's earlier negation of the Ghost's presence ("Nothing at all, yet all that is I see"). Hamlet had defined the Ghost as a "gracious figure" possessed of "piteous action" calling for mercy and kindness to Gertrude (3.4.104, 128). Gertrude's denial of the Ghost might then represent a negation of the visible presence of good. Her blindness seems to accord with the traditional Christian view of evil expressed by Thomas Aquinas: "Evil is the deprivation of good, and a privation is not a nature or real essence, it is a negation in a subject."[6] (Throughout the play, of course, Gertrude is unable to "see the Ghost"—to grasp the moral implications of her conduct, to perceive good as distinct from evil—any more than she has been able to "see" her two husbands—"Look here, upon this picture, and on this" [3.4.54].) Thus Hamlet's advice to Gertrude bitterly demonstrates the inversion of moral hierarchy in Claudian Denmark. In a world degraded by regicide and incest, good can be no more than the pale negation of a thriving evil. Hamlet's speech exhibits a despairing consciousness of ineradicable corruption.

Perhaps at some level of awareness Hamlet realizes that he is himself part of that corruption, even as in the speech itself he, the naysayer, is damned by what he denies. That is, if the poetic negative endows absence with presence, bringing "Not This" before us in the act of expressing it, then the naysayer must accept the guilt of his bringings-hither no less than the virtue of his forbiddings. "Nay, I had not known sin," Saint Paul said, "but by the law: for I had not known lust, except the law had said, 'Thou shalt not covet.' "[7] If the fate of the naysaying lawgiver is to forbid crimes into existence, that of the moralist executioner is to become guilty not merely of saying evil to deny it but of doing evil to destroy it. Hamlet is well aware that this paradox has spun a web for his own revengeful actions. Pointing to the dead Polonius, he acknowledges that—

> heaven hath pleased it so,
> To punish me with this, and this with me,
> That I must be their scourge and minister. (3.4.180–182).

The minister punishes sin without sacrificing his own virtue, but the scourge suffers the fate of the dyer's hand, which is subdued to what it works in—and Hamlet's hand is at work in the dark dyes of revenge and a leprous distilment that bathes the world of the court. In various ways he has sought sanctuary from the corruption about him, and increasingly within him, but by this time he has come to know the futility of that. Thrusting blindly through the arras, he has plunged himself most fatally into the contaminating milieu of action. And action, unfortunately, like things in nature, cannot be negated—it simply is.[8] Hamlet can renege on his word, as Polonius warns Ophelia ("Do not believe his vows; for they are brokers" [1.3.128]), but he cannot withdraw his swordthrust without discovering the blood of Polonius on his blade and the future madness and death of Ophelia on his conscience. From this perspective, Hamlet's advice to Gertrude illustrates the play-long process by which he comes to participate in the evil he seeks to purge.

Finally, perhaps most importantly, we can regard Hamlet's speech as symptomatic of the evaporation of differences and the consequent blurring of individual identities brought on by the acts of Claudius. If poetic negation is positive, then Not This exists on an equal footing with This. The absent is present, the denied affirmed, the forbidden consummated in the verbal act of negation itself. If the most radical demarcation of all, that between Being and Not-Being, is erased, how shall lesser distinctions stand? The discreteness of things cannot be guaranteed. Identities dissolve in a sea of undifferentiation, and the Uncertainty Principle is enthroned.

So it is in Demark at the opening of the play as we discover Horatio and the guards groping about in darkness in search of identities: "Who's there? . . . Stand, and unfold yourself . . . Bernardo? . . . Who's there? . . . What, is Horatio there?" Then as they await—not a Ghost, but something quite indefinable, "this thing" (21)—Bernardo speaks of how

> Last night of all,
> When yond same star that's westward from the pole
> Had made his course to illume that part of heaven
> Where now it burns, Marcellus and myself,
> The bell then beating one . . . (1.1.35–39)

And the Ghost enters, providing a visual predicate for Bernardo's uncompleted sentence and thereby creating a temporal indistinction between "last night" and this night. The Ghost itself, passing back and forth between otherworld and this-world, breaks down the borders between life and death and between fantasy and reason. Bernardo's casual "I have seen nothing" (22) is literalized by negative personification into "I have seen Nothing," as what Horatio first calls mere "fantasy" becomes "something more than fantasy" (23, 54). Then, in the interim between the two appearances of the Ghost, when Horatio speaks of the impending invasion by Fortinbras, the temporal merger of "last night" with this night in Bernardo's aborted sentence is reiterated as the incessant work of mobilization "does not divide the Sunday from the week" (76) and "doth make the night joint-labourer with the day" (78). The time is not merely out of joint, as Hamlet later complains, but appears to have no joint at all.

There is some suggestion here of a reversion to chaos, that vast sickening of enterprise so famously attributed by Ulysses to the "neglection of degree" (*Troilus*, 1.3.127).[9] Indeed, the confusion of Sunday and the rest of the week, of day and night, of otherworld and this-world, and of life and death suggests an undoing of the creative divisions that God labored over in *Genesis* as He gave form to the formless. God, surveying His work, beheld that "it was very good," but Hamlet, surveying Denmark, concludes that "It is not nor it cannot come to good" (1.2.158). God created Eden as Adam's world, but for Hamlet the world is "an unweeded garden/ That grows to seed" (1.2.135–136). And God established marriage, wherein a man "shall cleave unto his wife: and they shall be one flesh" (*Genesis* 2:24); but Hamlet declaims to Ophelia, "I say we will have no more marriage" (3.1.149–150), and ridicules Claudius as "My mother," for as he explains, "Father and mother is man and wife, man and wife is one flesh, and so, my mother" (4.3.55–56).

In these remarks Hamlet seems possessed of the Uncreating Word. But the discreations in Denmark are owing not to Hamlet but to Claudius, who in rising against his brother and sovereign, and overstepping the bounds between brother and sister-in-law, violates the medieval-renaissance concept of "Degree" by which men situate themselves within society and nature. The differences of Degree as they are ramified throughout life insure the identities of men and things. Without them, as Ulysses says, "Each thing meets/ In mere oppugnancy" (*Troilus*, 1.3.110–111). So it is with words too, especially in the unctuous first speech of Claudius:

> Therefore our sometime sister, now our queen,
> The imperial jointress to this warlike state,
> Have we, as 'twere with a defeated joy—
> With an auspicious and a dropping eye,
> With mirth in funeral and with dirge in marriage,
> In equal scale weighing delight and dole—
> Taken to wife. (1.2.8–14)

A funeral marked by mirth, it goes without saying, is not a funeral, nor is a marriage accompanied by dirge a marriage. In deference, if not to true feeling, at least to decorum, which distinguishes between occasions and styles, a deferment of the "o'erhasty marriage" might have preserved a trace of order and ceremony in what becomes instead a grotesque hybrid, a funeral-marriage.

The running together of funeral and marriage signals a breakdown of the borders between other entities. Hamlet the grieving son of the dead king discovers that he is not merely "cousin Hamlet" to the new king but "my son" as well, which is rather more kinship than he cares for: "A little more than kin, and less than kind" (1.2.65). Claudius is now his "uncle-father" and Gertrude his "aunt-mother" (2.2.376), a hyphenization of relations that leads to the total undifferentiation of Hamlet's reference to Claudius as "My mother" (4.3.55). This destruction and merger of identities leaves Hamlet longing in his first soliloquy for a corresponding dissolution of self, for a deliquescence of his corporeal being in a union with nothingness: "O, that this too too sullied flesh would melt,/ Thaw, and resolve itself into a dew" (1.2.129–130). All such repellent, hyphenized "unions" flow poisonously into the cup from which Gertrude drinks in the final scene and which Hamlet forces upon the already dying Claudius with the words, "Drink off this potion. Is thy union here?" (5.2.328). Hamlet's killing of Claudius is, in this context, an act of restorative destruction, an undoing of unions that came into existence not through the linking of like to like but through the disintegration of proper differences.

FOURTEEN

Amleth and Metaphor

CLAUDIUS' oxymorons, words united in mere oppugnancy, suggest that the process of undifferentiation acts as poisonously on language as it does on time, ritual, kinship, and the human body. Language, after all, is a major part of the world Hamlet inherits on his return from Wittenberg. Or, as it is more often put these days, the world Hamlet encounters is a product of the language he inherits. In either case we shall need to examine Hamlet's involvements with the negative as a response to a verbal milieu not of his own making. Let me advance upon that issue by indirections, taking a brief detour through Jutland to observe how Hamlet's Icelandic predecessor dealt with his similar difficulties.

In the story of Amleth told by Saxo Grammaticus, King Horwendil, successful in love and war, is waylaid and slain by his jealous brother Feng, who subsequently makes public confession of the murder but exculpates himself by claiming that he acted in defense of the gracious Gerutha, who had been threatened by her husband's "extremest hate." This lie gains instant credit at court, as lies will, Saxo remarks. Nevertheless, the grieving Amleth senses the truth, even without the aid of ghostly narratives, but fearing for his life he chooses "to feign dulness, and pretend an utter lack of wits. This cunning course not only concealed his intelligence but ensured his safety."[10] However, his silliness, as Saxo sometimes calls it, is suspected by some for wiliness, and is therefore put to the test. A major trial consists of placing a fair and willing female in Amleth's path. Feng's men assume that if Amleth yields to the flesh he will be of right mind, for passion is "too impetuous to be checked by cunning" (p. 105). Fortunately, however, Amleth is forewarned and so manages things that he appears witlessly chaste, though in actuality he has simply taken the girl to a secluded spot to demonstrate his concupiscent sanity. (The parallel incident in Shakespeare's play is a dramatic displacement of the myth, Hamlet responding to Ophelia in the Nunnery Scene as if she were a sexual lure of the grossest sort.)

More often, the tests of Amleth's madness take a verbal form. As he and Feng's men ride toward the waiting wench, for instance, a wolf crosses their path and (as Saxo continues):

> When his companions told him that a young colt had met him, he retorted, that in Feng's stud there were too few of that kind fighting. This was a gentle but witty fashion of invoking a curse upon his uncle's riches. When they averred that he had given a cunning answer, he answered that he had spoken deliberately: for he was loth to be thought prone to lying about any matter . . . and accordingly mingled craft and candour in such wise that, though his words did not lack truth, yet there was nothing to betoken the truth and betray how far his keenness went.
>
> Again, as he passed along the beach, his companions found the rudder of a ship which had been wrecked, and said they had discovered a huge knife. "This," said he, "was the right thing to carve such a huge ham"; by which he really meant the sea, to whose infinitude, he thought, this enormous rudder matched. Also, as they passed the sandhills, and bade him look at the meal, meaning the sand, he replied that it had been ground small by the hoary tempests of the ocean. His companions praising his answer, he said that he had spoken it wittingly. (pp. 107–8)

Here on three occasions Amleth is tested by a lie, a deliberate misnaming. These minor lies by Feng's men are extensions of the original lie issued by Feng himself about the murder of Horwendil, and Amleth's oft-expressed loathness to lie distinguishes him from his uncle. But Amleth's problem is how to preserve not merely the truth but also his life. If he did not care about the truth he could accept the lies of Feng's men, thereby proving his madness by his inability to tell wolves from colts, knives from rudders, and meal from sand. On the other hand, if he did not care about his life, he could simply reject their lies and insist on the right names for things. Caring about both, however, he must somehow incorporate the lifesaving lie and the soulsaving truth within the same utterance.

Amleth's solution is metaphor. The literal misnamings of Feng's men he accepts simply by converting them into figurations. For his purposes metaphor is an ideal instrument, because even in less desperate usage it is a form of rational madness that tells a true lie. From the standpoint of reason exemplified by Shakespeare's Duke Theseus or by Swift's Houyhnhnms, metaphor "says the thing that is not." A girl is not a red rose, a thought green, or all the world a stage. Whoever says they are must be either a madman or a liar. Or a metaphorist who knows that some truths can only be reached by a crooked road, and that to get there reason itself must sometimes trot with Tom o' Bedlam. Because Amleth can follow such a road, singing Tom's song as he goes, he keeps his two best possessions, his truth and his life.

FIFTEEN

ERASURES, POISON, AND NOTHING

IF AMLETH is saved by the negative, by the apparent non-sense and untruth of metaphor, so for a time is Hamlet, who employs negation to fashion the unspoken word and the unperformed act. I shall address this issue a bit further on in connection with Hamlet's wordplay and mad-play. For now let me glance at some simpler examples of Hamlet's employment of the negative.

Hamlet's special affinity for the negative is suggested when he first appears in Act 1, Scene 2. In the 46 lines he speaks before Horatio, Marcellus, and Bernardo enter to shift the focus, one finds 18 instances of the negatives "no," "not," "nor," and "nothing." That is, 39 percent of these lines contain negatives, as compared over the same scenic distance to 9.5 percent of Claudius' lines (9 negatives in 93 lines). To be sure, this does not tell us a great deal, except perhaps that the vocabulary of the grief-stricken is short on affirmatives, even when it seems agreeable ("Ay, madam, it is common" [1.2.74]), and that kings who come to the throne through secret murder are likely to be anxious yea-sayers even when using negative terms ("What wouldst thou beg, Laertes,/ That shall not be my offer, not thy asking?" [1.2.45–46]). Captious readers who are not deep in admiration of statistics, however, may be less impressed by the quantity of Hamlet's negatives than by the uses to which they are put.

A major use to which Hamlet puts the negative is erasure. Upon hearing the Ghost's revelations, for instance, he engages in a series of erasures, beginning with his erasure of all his former values and beliefs:

> Remember thee!
> Yea, from the table of my memory
> I'll wipe away all trivial fond records
> All saws of books, all forms, all pressures past,
> That youth and observation copied there,
> And thy commandment all alone shall live
> Within the book and volume of my brain,
> Unmixed with baser matter. (1.5.98–105)

What is erased, or wiped away, in this speech is made more explicit in Hamlet's explanation to Guildenstern in Act 2, Scene 2:

> I have of late—but wherefore I know not—lost all my mirth, forgone all custom of exercises; and indeed it goes so heavily with my disposition that this goodly frame, the earth, seems to me a sterile promontory; this most excellent canopy, the air, look you, this brave o'erhanging firmament, this majestical roof fretted with golden fire, why, it appeareth nothing to me but a foul and pestilent congregation of vapours. What a piece of work is a man! How noble in reason! How express and admirable in action! How like an angel in apprehension! How like a god! The beauty of the world! The paragon of animals! And yet, to me what is this quintessence of dust? Man delights not me—no, nor woman neither, though by your smiling you seem to say so.(2.2.296–311)

Having erased his schoolboy conceptions of reality and written "Remember thee!" in their place, Hamlet next erases the Ghost's revelations by consigning them to secrecy. He first appears about to tell his friends about Claudius ("There's never a villain dwelling in all Denmark") but then shies off into Polonius-like tautology ("But he's an arrant knave" [1.5.124–125]). The nonsense of his evasion and the rather frantic playfulness of his ensuing behavior, which prompts Horatio's bewildered "O day and night, but this is wondrous strange!" (165), suggest that Hamlet's former identity as the "expectancy and rose of the fair state" (3.1.155) is already being subjected to erasure in favor of an "antic" disposition." Then, having erased the Ghost's revelations, Hamlet next erases the appearance of the Ghost itself by repeatedly swearing his companions to silence about "what you have seen tonight" and "this that you have heard " (145, 154). And finally, in a speech that foreshadows his double-negative advice to

Gertrude, he pictures scenes in which his friends betray the fact that his antic behavior is assumed, and then erases them:

Here, as before, never, so help you mercy,
How strange or odd soe'er I bear myself—
As I perchance hereafter shall think meet
To put an antic disposition on
That you, at such times seeing me, never shall
With arms encumbered thus, or this headshake,
Or by pronouncing of some doubtful phrase,
As "Well, we know," or "We could, an if we would,"
Or "If we list to speak," or "There be, an if they might."
Or such ambiguous giving out, to note
That you know aught of me—this not to do,
So grace and mercy at your most need help you,
Swear. (1.5.169–181)

When Hamlet next appears—not in person but in Ophelia's story of his visit to her chamber—he has lapsed into an antic silence. His dumb show of disengagement erases his love for Ophelia—as does his behavior later in the Nunnery Scene, where he says "I did love you once" only to add "I loved you not" (3.1.116, 120)—and his dissheveled appearance erases the young prince who in former times was "the glass of fashion and the mould of form" (3.1.156). Hamlet graduates from silence with Ophelia to speech with Polonius, but even his verbal style has come under erasure. The fine blank verse he spoke before and immediately after the meeting with the Ghost has now gone the way of his once fine apparel. In its place—the verbal analogue to the visual Hamlet, hatless, doublet all unbraced, stockings fouled (2.1.75–78)—is an indecorous prose replete with indecent puns: the disordered style of his "antic disposition" as distinguished from the blank verse "sanity" of his soliloquies.

Hamlet's various erasures are incited by the Ghost's tale of murder and incest. From a verbal standpoint, however, the crucial revelation in the Ghost's tale is that Hamlet's father died from poison most curiously administered:

Upon my secure hour thy uncle stole,
With juice of cursed hebona in a vial,
And in the porches of mine ears did pour
The leprous distilment. (1.5.62–65)

Moreover, the poisonous contents of this vial, though perhaps literally used up on the sleeping king, remain metaphorically available afterwards for a second and similar application:

> It's given out that, sleeping in mine orchard,
> A serpent stung me. So the whole ear of Denmark
> Is by a forged process of my death
> Rankly abused. (1.5.36–39)

Thus the leprous distilment in the vial is transmuted into the lie that infects the ear of all Denmark.[11] An apt metaphor for the lie: verbal poison.[12] The body politic of Denmark, so infected, suffers throughout the play the lazarlike, vile, and loathsome effects that are writ small in the body of the dying king (1.5.64–73). Terence Hawkes traces the spread of this verbal contagion:

> As the supreme communicative instrument, then, the *sine qua non* of humanity, language at Elsinore has been seen to deteriorate progressively in the play, bringing with it a breakdown in exactly those personal relationships which underpin society and make man human. From the early formal blandishments of Claudius, polish concealing the truth of murder and usurpation, through the "indirections" of Polonius, which circle the truth but never approach it, down to Ophelia's ravings, the obstructive technicalities of the Gravediggers, and the "civilized" gibbering of Osric, mirroring the culture that has produced him, the process is an accelerating one of descent to a level of uncommunicative brutishness, where spying replaces talking and listening, and men become hunting and hunted animals.[13]

Thus when the Ghost admonishes Hamlet, "Taint not thy mind," he might with equal point have said, "Taint not thy speech." Amid the general contamination in Elsinore, the one seems as impossible as the other. Perhaps that is why, when we first see him, Hamlet stands for 63 lines wrapped in silence; why he ends his first soliloquy with the words "But break my heart, for I must hold my tongue" (1.2.159); and why in his third soliloquy he unexpectedly accuses himself not of an inability to act but of an inability to speak:

> A dull and muddy-mettled rascal, peak
> Like John-a-dreams, unpregnant of my cause,
> And can say nothing. (2.2.566–569)

Perhaps that is also why, on the one occasion when he tries to speak the language of honest feeling—in his "Give me that man/ This is not passion's slave" speech to Horatio—he breaks off abruptly, muttering "Something too much of this" (3.2.73).

In Act 1, Scene 2, however, Hamlet does not yet know about the poisoning of his father. His disapproving silence there is a reaction to his father's death but also to the incestuous liaison of his mother and Claudius, which in its own way is as corruptive of words as the King's unroyal lies. For if Gertrude will lie with Claudius as readily as with Hamlet's father, then words themselves will lie with any meaning. Claudius lies to the kingdom, Gertrude lies with Claudius, thereby making a lie of her former marriage: "Such an act/ That . . ./ . . . makes marriage-vows/ As false as dicers' oaths" (3.4.41–46). Here is the incest of language, when words, truth, and lies all foully embrace, "honeying and making love/ Over the nasty sty" of the world as Hamlet now sees it (3.4.95–96). Between incestuous sheets Hamlet discovers the unnatural couplings of Claudius' oxymorons and the perverse unions of funeral and marriage, uncle and father, aunt and mother, and nephew and son. There too, it seems, are engendered his own incestuous abuses of words: his puns, riddles, equivocations.[14]

If speech, the medium of truth, is tainted, then truth itself can only withdraw into silence, as Cordelia so painfully knows. When finally compelled to speak, Cordelia, who would say nothing, does the next best thing: she says "Nothing" (*Lear,* 1.1.89). Hamlet, in a rather similar situation, also says nothing at first, keeping a black silence amid the urbanities of court business, until his mother asks him why if death is common it "seems" so particular with him. Then he too says "Nothing," at somewhat greater length than Cordelia:

> Seems, madam! Nay, it is. I know not "seems."
> 'Tis not alone my inky cloak, good mother,
> Nor customary suits of solemn black,
> Nor windy suspiration of forced breath,
> No, nor the fruitful river in the eye,
> Nor the dejected haviour of the visage,
> Together with all forms, moods, shows of grief,
> That can denote me truly. These indeed seem,
> For they are actions that a man might play;
> But I have that within which passeth show,
> These but the trappings and the suits of woe. (1.2.75–85)

As the eight conspicuous negatives suggest, this is another erasure speech along the general lines of his admonitions to his friends and mother about his "antic disposition" and madness "in craft." In those speeches, however, he sought to erase hypothetical future acts of betrayal, whereas here he seeks to erase present appearances, all "forms, moods, shows of grief." Not that his display of grief is false. Far from it. But amid the general debasement of signs in Elsinore—the poisoning of truth, the promiscuity of words, the reduction of "sweet religion" to a "rhapsody of words" (3.4.48–49)—all symbols, ceremonies, forms, gestures, outward signs and shows are drained of meaning, his own as well. Hamlet's grief is betrayed by courtly seeming, by the enveloping falseness of a society in which true feeling must, in which it can *only*, "appear."

The individual truth of Hamlet's grief is thus erased through its absorption by a world of surface show. With this discrediting of the surface, what remains is an undefined inner presence, "that within which passeth show." The inner meaning somehow exceeds its outer sign ("passeth" = surpasses), but what that surplusage may be is left to our and Claudius' imagination. Claudius would no doubt like to take Hamlet's lines as meaning no more than "My grief is too great to convey." But as Hamlet's first soliloquy indicates, and as his exclamation when told of his uncle's villainies darkly confirms ("O my prophetic soul!/ Mine uncle?" [1.5.40–41]), Hamlet's inner self even at this early point holds not merely grief but also ghostly imaginings and fierce resentments too great for public expression Denmark.

In his use of negation in this speech Hamlet in effect says "Nothing," only his word is "Seems." In a sense he becomes Nothing, for by negating his surface appearances he erases his public identity in Elsinore. That is surely a consummation devoutly to be wished, to retreat into the fortress of self and keep the outer world at bay. And yet of course Hamlet remains within that world, with his suit of solemn black and all his actions that a man might play. How else? Grief, like love, has no private language. The word *love* in Lear's court and the symbols of grief in Elsinore may be equally degraded, yet Cordelia must still say "I love you" and Hamlet adopt the trappings and the suits of woe. By calling on the negative, however, Hamlet has come as close as he can to erasing his public self and retreating into an inner world of unspoken meaning.[15]

SIXTEEN

SILENCE, SOLILOQUY, COURT SPEECH, NOISE

LET ME turn now toward Hamlet's wordplay, and especially to his use of the pun, that fatal Cleopatra who proves as seductive amid the fens and bogs of Denmark as she was along the Nile.[16] For Hamlet is the master quibbler in a play that has, as Molly Mahood observes, "more quibbles than any other of Shakespeare's tragedies."[17] Thus he begins his part in the play with a pun, "A little more than kin, and less than kind" (1.2.65), and concludes it with another, "The rest is silence" (5.2.360)—"rest" here meaning both remainder, of which Hamlet has none, and repose, tranquillity, cessation of movement, all the meanings once contained in his pacification of the Ghost ("Rest, rest, perturbed spirit!" [1.5.183]) and now so fittingly applied to himself.

Hamlet plays on and with words for various reasons. The reason usually given focuses on the self-protective function of his assumed madness. As Wolfgang Clemen puts it:

> Hamlet needs images of his "antic disposition." He would betray himself if he used open, direct language. Hence he must speak ambiguously and cloak his real meaning under quibbles and puns, images and parables.[18]

This is surely an immediate pragmatic reason both for Hamlet's punning and for Amleth's metaphorizing. But perhaps a glance at the nature of the pun, and of wordplay in general, will suggest other reasons for its prominence in Hamlet's speech. To gain a somewhat broader perspective, however, let me first place Hamlet's wordplay within a larger context of sound, in fact at the center of a continuum of discourse in Elsinore. That is, sound and silence in Elsinore might be divided up along these lines:

Silence – Soliloquy & Aside – Wordplay – Court Speech – Noise

This is not of course a temporal ordering (a plot generated by silence and climaxed by noise) but a conceptual one. Even in a temporal plot, however, *silence* would play a significant role. We encounter it in the opening speech, with the Ghost's refusal to speak to Horatio and the others (" 'Tis gone, and will not answer"); in the second scene, with Hamlet's dark wordlessness while Claudius holds forth ("But break my heart, for I must hold my tongue"); in the second Ghost scene, with Hamlet's insistence on secrecy ("Never to speak of this that you have heard"); in Act 2, Scene 1, with Hamlet's dumb show of estrangement from Ophelia; in Act 3, with the Dumb Show itself; in the Prayer Scene, with the double silences of Hamlet and Claudius; and so on to the murderous silence of the duel and the final stillness of multiple death. In Hamlet's own experience silence is alpha and omega. He begins by enacting silence in the Council Scene and ends by announcing it in the Duel Scene ("The rest is silence"). Silence is the matrix from which the play issues, the inexpressible country from which the Ghost comes like a whispered word to make us realize that beneath the chatter and clamor of the court there was, all along, a more meaningful zone of silence and secrecy. The Ghost's meanings rise from silence to the surface of speech, are conveyed into Hamlet's ears like the contents of Claudius' vial, and are consigned again to silence. Hamlet is now the festering ulcer of silence, a hidden imposthume of foul and inexpressible meanings.

If silence in *Hamlet* could be defined as meaning without a voice, then *noise* at the opposite extreme could be called a voice without meaning. Extremes, as Coleridge liked to note, often meet. In the present case they meet at the point of incomprehensibility. Thus Ophelia asks when the Dumb Show is over, "Will they tell us what this show meant?" (3.2.141), and Hamlet in his advice to the Players inveighs against "inexplicable dumb shows and noise" (3.2.13), the adjective, I take it, modifying both nouns. Voiceless meanings and meaningless voices are equally incomprehensible.

Perhaps the most raucous noise in Elsinore is the repeated cacophany that rises from the banquet hall even to the castle platform to announce, as Hamlet says, that—

> The King doth wake tonight and takes his rouse,
> Keeps wassails, and the swaggering up-spring reels;
> And, as he drains his draughts of Rhenish down,
> The kettle-drum and trumpet thus bray out
> The triumph of his pledge. (1.4.8–12)

Hamlet complains that this "heavy-headed revel," which breaks down "the pales and forts of reason" in individual men, has become in Denmark a corrupting national trait. As such it is amply illustrated. Claudius had earlier promised to honor Hamlet's filial obedience (in not returning to Wittenberg) with cannonades and toasts:

> No jocund health that Denmark drinks today
> But the great cannon to the clouds shall tell,
> And the King's rouse the heaven shall bruit again,
> Respeaking earthly thunder. (1.2.125–128)

These hubristic challenges to the heavens are repeated again during the Duel Scene at the King's order:

> Give me the cups,
> And let the kettle to the trumpet speak,
> The trumpet to the cannoneer without,
> The cannons to the heavens, the heaven to earth,
> "Now the King drinks to Hamlet." (5.2.272–276)

These various brayings-out of cannon, drum, and trumpet assault the ear as Rhenish does the reason and as drunkenness itself does the national reputation. They suggest an untuning of the natural harmonies of mind and state, a brassy intoxication of sound, and a recurrent example of "That monster, custom, who all sense doth eat" (3.4.168 ff.; cf. 1.4.12 and 3.4.38).[19]

In counterpart to this public noise is the soft utterance of the mad Ophelia, equally bereft of sense, as the unnamed Gentleman says:

> Her speech is nothing,
> Yet the unshaped use of it doth move
> The hearers to collection. They yawn at it
> And botch the words up fit to their own thoughts,

> Which, as her winks and nods and gestures yield them,
> Indeed would make one think there would be thought,
> Though nothing sure, yet much unhappily. (4.5.7–13)

Since inexplicable speech is no more than noise, it is fitting that Ophelia's next entrance be preceded by the stage direction, "A *noise within:* 'Let her come in!'" followed by Laertes' "How now! what noise is that?" (4.5.154–155).

That Ophelia's gentle speech has become nothing is sad, but not inconsistent with the fact that she mourns a father who was given to diagnoses and definitions of the following sort:

> Your noble son is mad.
> Mad call I it; for, to define true madness,
> What is't but to be nothing else but mad? (2.2.92–94)

The triumph of sound over sense that so often occurs when Polonius speaks is carried to extremes in his children's responses to his death. Thus the madness of Ophelia in 4.5 is paralleled in the same scene by the derangements attending her brother's entrance, an entrance that, like hers at line 154, is preceded by the stage direction "A *noise within*" and by the Queen's startled "Alack, what noise is this?" (4.5.97). The inward assault upon the pales and forts of Ophelia's reason has its outward semblance in the assault upon political order by Laertes:

> The ocean, overpeering of his list,
> Eats not the flats with more impitious haste
> Than young Laertes in a riotous head,
> O'erbears your officers. The rabble call him lord;
> And, as the world were now but to begin,
> Antiquity forgot, custom not known,
> The ratifiers and props of every word,
> They cry, "Choose we! Laertes shall be king!"
> Caps, hands, and tongues applaud it to the clouds,
> "Laertes shall be king, Laertes king!" (4.5.102–111)

The image of the ocean eating the flats is a recurrent Shakespearean symbol of the erasure of distinctions and a return to chaos. "Take but degree away," as Ulysses says, and (among other things),

> The bounded waters
> Should lift their bosoms higher than the shores
> And make a sop of all this solid globe. (*Troilus*, 1.3.109 ff.)

Or with "antiquity forgot" and "custom not known"—

> everything includes itself in power,
> Power into will, will into appetite,
> And appetite, an universal wolf,
> So doubly seconded with will and power,
> Must make perforce an universal prey,
> And last eat up himself. (*Troilus*, 1.3.119–124)

With the "ratifies and props of every word" so undermined, "the consequent decay of temperate and intelligible speech augurs the unmaking of civilization: 'As the world were now but to begin.' "[20]

The ubiquitous drums and cannon of Claudius invite all speech toward unintelligible excess. Hamlet complains of actors who "tear a passion to tatters" and "split the ears of the groundlings" (3.2.10–12). Yet he himself "out-herods Herod" when, thinking of Claudius, he falls "a-cursing, like a very drab,/ A scullion" (2.2.587–588); when he raises verbal daggers against his mother, who cries "Ay me, what act,/ That roars so loud and thunders in the index?" (3.4.52–53); and when he "prates" competitively with Laertes in the grave of Ophelia ("Nay, an thou'lt mouth,/ I'll rant as well as thou" [5.1.283–284]).

The insurrection that would send Laertes to the Danish throne is the mutinous corollary of the secret usurpation by which Claudius achieved his own kingship. Murder in the orchard makes way for riot in the state, as King Claudius ushers in the would-be King Laertes. What Rosencrantz calls "the cease of majesty" took place when Claudius killed his brother and stole his crown; and that cessation "Dies not alone, but, like a gulf, doth draw/ What's near it with it" (3.3.16–17). Thus the imposthume hidden within the realm of Claudius erupts to take public form as Laertes' mutiny. But within Claudius himself the imposthume is kept well concealed. His devious ascension is even more insidiously destructive of order than Laertes' mad rush for kingship—but only Hamlet knows that, Claudius having kept up political appearances so well. Ulysses speaks of the dangers of coupling "will and power," but Claudius takes care not to appear willful. Indeed he derives part of his power from his obedience to the wills of others. Thus his kingship is a product of the votes with which he stole the election from Hamlet, and his marriage is deferentially in accord with "Your better wisdoms, which have freely gone/ With this affair along" (1.2.15–16).

If the ascension of Claudius is a subtilization of Laertes' grab for power, so his speech is a subtilization of the meaningless noise of Laertes' mob. Moving to the left on the continuum of discourse, that is, we come to *Court Speech:* the language of the King and Queen, Polonius, Osric, Laertes, and Rosencrantz and Guildenstern. Court speech is naturally dominated by the King. And the speech of the King has at its corrupted core the Lie, which by loosing the bond of meaning between words and truth thrusts discourse in the direction of Noise. The King's lies are not restricted to the "forged process" of Old Hamlet's death; they invade much of what he says, especially during his first appearance (1.2), where the unguent oxymorons of his opening speech seek to close the deep social wound caused by the late king's death and his wife's remarriage. But the most doleful marriage in this speech is verbal. The rhetorical union of such hopelessly irreconcilable experiences as funeral and marriage serves only to divorce words from reality. Here, and in the King's remarks to Hamlet in this scene—

> But now, my cousin Hamlet, and my son . . . (65)
>
> 'Tis sweet and commendable in your nature, Hamlet . . . (87)
>
> Here in the cheer and comfort of our eye . . . (116)
>
> Why, 'tis a loving and and a fair reply . . . (121)
>
> This gentle and unforced accord of Hamlet
> Sits smiling to my heart . . . (123–124)

—Shakespeare demonstrates that when the lie comes to court it takes the form of hypocrisy, the verbal lubricant that reduces the friction between false words and true meanings to an amiable hiss.

It is not necessary to catalog the tautologies of Polonius, the airy ostentation of Osric, the vacuities of Rosencrantz and Guildenstern, and other forms of verbal inflation around Claudius to justify locating Court Speech as a close neighbor of Noise on the continuum of Danish discourse. It is Hamlet's recoil from such nonlanguage in part that sends him, like Cordelia, into despairing silence and lengthy speeches of self-erasure. Given the contaminations of Court Speech the only sure means of satisfying the Ghost's hypothetical "Taint not thy speech" is by not speaking at all. Absolute silence, however, is difficult to maintain, especially for a character in a play. So Hamlet takes a step in the direction of public speech by responding to Claudius' "But now, my cousin Hamlet, and my son—" with a statement of his true feelings: "A little more than kin, and less than kind." But he does so in an aside.

One can think of *Asides* as brief soliloquies or of *Soliloquies* as sustained asides. In either case they are self-negating forms of utterance: speeches that unsay themselves by the articulation of silence. Thus when Hamlet ends his first soliloquy with the words "But break my heart, for I must hold my tongue" we may be surprised on reflection to realize that that is just what he has done, hold his tongue. The words we have just heard were not uttered. Asides and soliloquies are duplicitous creatures capable of a chameleonlike reversibility between speech and silence. Insofar as they are unspoken, they are a redundant category of discourse, an appendage of silence itself, into which (if we blink our eyes) they disappear, just as at the other end of the continuum Court Speech keeps shading off into pure Noise. Insofar as they are spoken, however, they represent a gesture away from silence toward the verbal world of the court, and hence from unvoiced meaning at one extreme toward meaningless voices at the other.

SEVENTEEN

At the Center: Wordplay as Negation and Erasure

SILENCE, EVEN an articulate silence, is not sufficient for Hamlet's cause. He must speak aloud in Elsinore, and so move into an area of discourse lying between Silence and Soliloquy on one side and Court Speech and Noise on the other, a region of wordplay dominated by the pun. And why the pun? In part because like all figurative language it takes its source in the negative. Whereas literal speech, aiming at an unambiguous transfer of information, seeks an identity between what is said and what is meant, figurative speech is not figurative unless what is said differs from what is meant. Wordplay intrudes the eraser of negation between signifiers and signifieds. When Hamlet replies to Claudius' question, "How is it that the clouds still hang on you?" by saying "Not so, my lord; I am too much in the sun," his quibble eclipses the solar meaning of the last word and writes the filial meaning in its place. At least it does so for the audience. Claudius fails to reply to Hamlet's substitute meaning (though of course the actor may wince or scowl if he choose), either because he does not register that meaning or because he prefers to ignore it. In either case his lack of response accords with the fact that Hamlet's pun is homophonic. Homophonic puns are created by an accident in language, an unintentional rendezvous of disparate meanings. Two words with no semantic connection whatsoever, like *sun* and *son*, happened on some dark night in their linguistic past to wander into the same phonemic lodgings, which they have shared ever since.

Not so with the puns in Hamlet's opening line, "A little more than kin, and less than kind," which are polysemous instead of homophonic. That is, there is a fairly close semantic kinship among the tribe of meanings engendered by *kin* and *kind:* family, relations, class, natural, well-born, well-disposed. The degree of negation in polysemous puns, the gap between what is said and what is meant, is smaller than that in homophonic puns. In Hamlet's opening line that gap is entirely too narrow for courtly comfort, and as a result editors have quite properly termed the remark an aside. The King can dismiss homophonic puns as accidental, an antic disposition of meanings—as when Polonius says "How pregnant sometimes his replies are! a happiness that often madness hits on, which reason and sanity could not so prosperously be delivered of" (2.2.208–211)—but polysemous insults are too conspicuously clever to be unintentional.

Wordplay is not merely a means of self-protection for Hamlet, a concocted symptom of his "antic disposition." By offering him the eraser of negation, wordplay also enables Hamlet to purge words of their pretence to meaning—their bland and unctuous vacuity, their false shows and received smugness, their urbane noise—and invest them with meanings of his own. This dissociative process is illustrated when Claudius asks "How fares our cousin Hamlet?" and is answered,

> Excellent, in faith—of the chameleon's dish. I eat the air, promise-crammed. You cannot feed capons so.

To which the King:

> I have nothing with this answer, Hamlet; these words are not mine
>
> (3.2.91–95)

The King is right; these words are no longer his. Hamlet has erased his words and written in his own, which have identical sounds. He first confiscates the word "fares" (= "does") and replaces it with "fares" (= "feeds"), and then follows with a substitution of "heir" for "air." Crammed with the windy promises of the King ("You are the most immediate to our throne" [1.2.109]), Hamlet the rightful heir to his father's throne is, like the supposedly air-eating chameleon, obliged to keep his royalty in self-protective concealment, in part because like the capon, the rooster whose nature and destiny have been altered by castration (= the "un-crowning" of Hamlet), he may find himself invited to a dinner where he does not feed but is fed upon. With this answer the King can either make nothing or prefers not to, since what it implies is not merely a partially disclosed second meaning but also a partially disclosed second self. Within or behind the surface utterance of the mad and ineffectual court jester Hamlet, from the far side of his puns, speaks an ominously sane and perhaps lethal would-be-king. That is a Hamlet whose presence the King-that-is both suspects and fears, half wants to confirm, half wants to forget. Hamlet's puns, instruments of concealed disclosure, are precisely calculated to keep Claudius suspended between suspicion and safety.

Hamlet's wordplay is itself suspended between Soliloquy and Court Speech and, at a further stretch, between Silence and Noise. Although his puns are very rarely heard by anyone except himself (and of course us the audience), they are not entirely silent. But neither are they quite audible. They issue, like his soliloquies, from silence and rise to a point tantalizingly close to the surface of speech, achieving a sort of half-whisper that causes his hearers to turn back puzzled, as though to query "My lord . . . ?"—in response to which Hamlet would merely smile and raise his eyebrows in innocence. In such cases the unapprehended secondary meanings of Hamlet's wordplay become split-second soliloquies, flashes of intelligence uttered and unuttered in the same instant. Thus a form of cross-erasure occurs: Hamlet's wordplay erases courtly inanities wittingly, and the court world erases Hamletic ironies unwittingly.

This is a rather sour perversion of the pun. In an ideal society of discourse, perhaps at the end of one of Shakespeare's festive comedies or romances, puns would negotiate between silence and noise, and between soliloquy and court speech. Wedding private meanings to public utterance, wordplay would unite individual and society in meaningful dialogue. But such dialogue is no longer possible in Denmark, and Hamlet himself has declared an end to marriages (3.1.155). The perfect marriage, courted in sonnet and comedy, is the marriage of true minds that occurs when the truth of the feelings is truly said and truly heard. Words then fashion a union of minds that is sacramentally bonded by the "speech act" of the priest, "I pronounce that they bee man and wyfe together." Such a pronouncement is valid only if it issues from proper authority. When Owen Glendower announces, "I can call spirits from the vasty deep," Harry Hotspur takes leave to doubt the Welshman's calling: "Why, so can I, or so can any man;/ But will they come when you do call for them?" (*1 Henry IV*, 3.1.51–53). In *Hamlet* spirits like the Ghost arrive uncalled for, with news that undermines all authority, making a mockery of kingship and marriage. Even the King now, ultimate authority in the state, cannot communicate with spirits ("My words fly up, my thoughts remain below./ Words without thoughts never to heaven go" [3.3.97–987]), and spying has replaced dialogue as the means of communication between men. In the comedies the bandying of wordplay between lovers is often but a faintly disguised substitute for sexual play: a coupling of meanings, an impregnation of the word. But the marrying procreant word of comedy gives way in Hamlet's practice to the divorcing barren word: "To a nunnery, go." There are no marriages in either convents or brothels, and no children either. The meanings of "nunnery" multiply emptily.

At times of course Hamlet's puns have a comic effect. But it is the audience that laughs, not the other characters, who are given instead to baffled shakings of the head and meaningful pointings to the head. Not that anyone expects wordplay in *Hamlet* to exhibit the happy gregariousness it does in comedy. The comedic sense of community must yield here to tragic divisions. Thus the process of cross-erasure splits the pun's meanings down the middle, leaving Hamlet isolated with his ironies and his listeners penned in confusion. Taken literally, wordplay is indistinguishable from madness, as Hamlet's literal-minded listeners are ready to believe as they squint warily at him:

> Yet he knew me not at first; he said I was a fishmonger. (2.2.188)
>
> O, what a noble mind is here o'erthrown. (3.1.153)
>
> I have nothing to do with this answer, Hamlet.(3.2.94)
>
> Good my lord, put your discourse into some frame, and start not so wildly from my affair. (3.2.307–308)
>
> Take you me for a sponge, my lord? (4.2.15)
>
> What dost thou mean by this? (4.3.30)

As these blank responses indicate, Hamlet's wordplay registers in the public ear not much differently than the "unshaped use" of Ophelia's speech, which moves her "hearers to collection" (4.5.8–10). To be misunderstood and taken for mad is of course one purpose of Hamlet's "antic disposition" and whirling words. As his Icelandic predecessor knew, bouncing figurative truths off literal minds has its advantages. It keeps Hamlet alive, and it enables him to touch verbal pitch without defilement. But this preservation of life and purity comes at the high cost of not being heard, of becoming in a real sense a John-a-dreams who "can say nothing" (2.2.568). If the truth is to emerge from silence it must take a form that permits its meanings to be grasped. Early in the play, however, Hamlet has not found a voice that will make the truth audible in Denmark, nor has he found how to kill the King. These two endeavors, to tell the truth and to kill the King, are complementary and in some respects parallel.

EIGHTEEN

Wordplay, Madplay, Inner-Play

A GREAT many truths interest Hamlet, but the ones he must find a voice for are of two kinds. There are the fundamental truths imparted to him by the ghost: the secret facts about his father's murder and his mother's faithless remarriage. Then there are the truths about the human condition that seem entailed by these primary facts: all the melancholy distresses that Hamlet registers about a mutable world in which the cosmetics of lies and false seeming conceal the moral ugliness of evil. These latter truths Hamlet both tells and untells in his wordplay, engaging in a verbal balancing act between silence and noise that is paralleled by his assuming an antic disposition that lies between inaction and action.

Regarding action: Hamlet's predicament is caused by his need to reconcile the Ghost's command, "Revenge his foul and most unnatural murder," and its perplexing codicil, "But, howsomever thou pursues this act,/ Taint not thy mind" (1.5.26, 5–86). How to act within a tainted world without becoming tainted oneself? How, on the other hand, not to act without suffering the stigma of filial betrayal? Hamlet's solution for the moment is to take refuge in the cleft between action and inaction. He does not act but instead "acts," that is, plays mad, which places his behavior in *epochē,* zoned off from the world of pragmatic affairs in which action and inaction have meaning. Hamlet's antic disposition thus becomes a form of inactive action, to which the verbal analogy is the unvoiced speech of his wordplay. And inasmuch as *play* is largely an end in itself rather than a means to other ends, so Hamlet's wordplay fails to communicate the truth in Denmark and his madplay fails to kill the King.

Part of the reason for this suspension of action and speech is that Hamlet is not altogether persuaded that the Ghost's truths are truths, or even that the Ghost is merely a ghost.[21]

> The spirit that I have seen
> May be a devil, and the devil hath power
> To assume a pleasing shape; yea, and perhaps
> Out of my weakness and my melancholy,
> As he is very potent with such spirits
> Abuses me to damn me. (2.2.599–604)

The way to resolve his uncertainties about the Ghost, about Claudius, and about his own imagination is, he decides, theatrical: "The play's the thing/ Wherein I'll catch the conscience of the King" (2.2.605–606). The way to resolve his verbal and actional dilemmas is also theatrical. In rewriting and presenting "The Murder of Gonzago" he will accept his own advice to the players and "Suit the action to the word, the word to the action" (3.2.17–18). The silent truths about murder in the orchard and remarriage in haste will be made manifest to eyes and ears—

> For murder, though it have no tongue, will speak
> With most miraculous organ. (2.2.594–595)

Silence thus acquires a voice to make its meanings heard, if not by everyone, at least by Claudius. Like Hamlet's wordplay and madplay, "The Murder of Gonzago" in its original form was presumably an artistic end in itself. As revised by Hamlet, however, it becomes "The Mousetrap." Dramatic art is transformed into pragmatic instrument, a weapon in the real world. Suited mutually to one another to comprise the dramatic ensemble, words now act and acts speak, so that in one concerted motion the truth is conveyed to Claudius with the impact of a sword thrust.

NINETEEN

Praying King, Intruding Fool

THE SUCCESS of "The Mousetrap" seems to mark an end to Hamlet's own forms of play. He is now sent, licensed to act, into the real world. Not that he abandons wordplay entirely. He can still addle Polonius by showing him a visual pun, a polymorphous cloud inhabited by a menagerie of camels, weasels, and whales; and he can tell Rosencrantz and Guildenstern that though they can fret him they cannot play upon him as upon a recorder. But his puns now take on a sharper edge, as when he speaks of the risks to the King should Hamlet himself be called on to purge him of his "choler," and they are interspersed with expressions of icy enmity, to Rosencrantz ("Have you any further trade with us?" [3.2.332]) and to Guildenstern ("It is as easy as lying" [3.2.356]). This honing of his verbal weaponry prepares for his employment of words as "daggers" in the Closet Scene.

Before Hamlet rounds fully on his mother, however, he encounters Claudius and Polonius, one at prayer, the other behind the arras. These two meetings seem designed to demonstrate the depth of evil in the real world and the diabolical complexity of *acting* once one has abandoned "acting." Let us glance first at the Prayer Scene, an episode that is virtually a dumb show insofar as it consists almost entirely of soliloquies. This produces a curious situation in which the context of reality, the fact that the two men cannot hear one another, negates all that is said, and yet what is said is there to be negated. Similarly, the soliloquy of Claudius displays the overriding presence of an evil that should be, in orthodox terms, but the absence of good. The indelibility of evil is most apparent in Claudius' longing for an erasure of his "offence" that will leave it still intact: "May one be pardoned and retain the offence?" (3.3.56). In the end unrepentance negates repentance, and his words, bereft of true feeling, become empty noises that fall far short of heavenly hearing.

Hamlet's soliloquy has its share of evil as well:

> And am I then revenged,
> To take him in the purging of his soul,
> When he is fit and seasoned for his passage?
> No!
> Up, sword, and know thou a more horrid hent.
> When he is drunk asleep, or in his rage,
> Or in the incestuous pleasure of his bed,
> At game a-swearing, or about some act
> That has no relish of salvation in't—
> Then trip him, that his heels may kick at heaven,
> And that his soul may be as damned and black
> As hell, whereto it goes. (3.3.84–95)

Hamlet's explosive "No!"—so prominently isolated—is a negation of evil, if we apply to his would-have-been murder the standards of the Ghost when it called "Murder most foul, as in the best it is" (1.5.27). This is not, however, the orthodox negation of evil by good but a negation of evil by a greater evil in prospect, a revenge that kills the soul as well as the body.

If evil refuses to be the negation of good here, so inaction refuses to be the negation of action. Hamlet's inaction, our sense of him as doing nothing at this point, is a product of his soliloquy. That is, if we subtracted his soliloquy from this scene, we should be left with a dumb show of silent actions; and actions are as devoid of the negative as conditions in nature. Hamlet may *say* "I will not kill him now but later," but what he *does* is to stand, to unsheathe his sword, to sheathe it again, and to walk away (with whatever minor embellishments the actor provides). Whatever we may hear, in other words, we *see* only positive acts. From these acts we may infer negation and a failure to act by thinking along the lines of the soliloquy itself, by verbalizing its implicit "not." But the soliloquy itself does that for us, expanding the space between Hamlet's unsheathing and sheathing of his sword and filling that space with a "No!" that is complexly articulated. Thus his inaction is not a lack, an absence of action, but a vicious presence. It does not exist because he has failed to seize an occasion to act but because he seizes an occasion not-to-act. Inaction is an iniquitous positive alternative. Hamlet's "Up, sword" is, as an "unthrust," more cruel than any thrust he might have made at Claudius.

All of these inversions submit to a further wrench when we discover that the kneeling Claudius has been as vulnerable to damnation as Hamlet could have wished. Hamlet's revengeful "imaginations" have been "as foul/ As Vulcan's stithy" (3.2.82–83), and all to no end. In the following scene Hamlet does not fare a great deal better.

When Claudius is at prayer, Hamlet is led astray by silence. When Polonius is behind the arras, Hamlet is led astray by noise, by the racket put up by the terrified old man, "What ho! help!" (3.4.23) The juxtaposition of the Prayer Scene and the Closet Scene offers a clear demonstration that Silence and Noise are equally, and for Hamlet disastrously, inexplicable. Had Claudius prayed aloud, he would be dead; had Polonius kept silent, he would be alive. Betrayed by the sight of Claudius seemingly at his prayers, Hamlet departs without knowing what he failed to do. Betrayed by the sounds of Polonius behind the arras, he acts without knowing what he does.

> *Queen.* O me, what has thou done?
> *Ham.* Nay, I know not.
> Is it the King? (3.4.26–27)

On the one hand an excess of thought leaves the true Claudius unkilled, and on the other hand an insufficiency of thought leaves the false Claudius dead.

The linking of these two episodes by parallels and contrasts throws into relief the self-defeating nature of both inaction and action, at least when undertaken in the present Hamletic manner. Having emerged from his own sanctuary of "seeming" (wordplay and madplay), Hamlet finds himself spun about in Claudius' castle of mirrors, a world of "real" seeming where the appearance of prayer can hide the truth as effectively as an arras can conceal the identity of an old meddler. Despite the fact that he has tainted his mind in the Prayer Scene and his sword in the

Closet Scene, Hamlet exhibits an almost hubristic confidence in the rightness of his actions. Thus he turns from his "rash and bloody" killing of Polonius to a rash and very nearly bloody attack upon his mother, a sustained invective so vicious that she is almost physically beaten down by it: "O, speak to me no more!/ These words like daggers enter in mine ears" (3.4.97–98). So close indeed does Hamlet's verbal abuse approach the physical that the arrival of the Ghost seems timed precisely to remind him that he was forbidden to "contrive/ Against [his] mother aught" (1.5.86–87).

Hamlet's errors of revenge present us with a coalescence of dramatic and divine irony. His presumption in taking upon himself the damnation of Claudius and the death of whoever stood behind the arras is silently rebuked by higher authorities—by Shakespeare the fashioner of tragic plots and by God the fashioner of providential plots. Shakespeare's plot, not Hamlet, will select the time of the King's death, and the divine plot of providence will determine the fate of his soul. Hamlet's fault, it appears, is not in choosing either to act or not to act, since both fail, or in thinking too precisely or imprecisely on the event, since both fail, but in arrogating to himself supreme authority on each occasion. In an understandable release of aggression following the revelations of "The Mousetrap," Hamlet has proceeded to two extremes of revenge and, somewhat like Oedipus proceeding from the revelations of the oracle to the ironies of the crossroads, suffers a *peripeteia* in each instance.

These two episodes set the stage for, and are retroactively illuminated by, Hamlet's change of attitude when he returns from his sea-voyage in Act 5. His references there to a "divinity that shapes our ends," to a "special providence" that presumably concerns itself with princes no less than with sparrows, and to a kind of spiritual poisedness between inaction and action (a "readiness" that is "all") imply his abdication from the kind of presumptuous revengefulness that failed so abjectly in the Prayer and Closet Scenes. This new acceptance of himself as playing a role in a larger plot prepares him and us for a revenge that does not issue from an overcunning scheme to damn the King's soul or from a rash act performed in willful ignorance but that emerges from the evolving situation itself. Such a revenge, latent in the occasion, is not initiated by Hamlet but accepted; it is less an action than a reaction.[22]

TWENTY

HAMLET AS PUN

LET ME return to Hamlet's wordplay. It has served in part to divide Hamlet from the court by erasing court meanings and substituting Hamlet's ironies, which the court in turn erases by unhearing, so that hero and society are sealed off from one another as hermetically as the literal and figurative meanings in Hamlet's wordplay. One virtue of Hamlet's wordplay is that, as a form of not-saying (paralleling his antic disposition as a form of not-acting), it has shielded him against verbal contamination in Elsinore. One drawback, however, is that as a form of not-saying it has not said what needs saying in Denmark, the truth, just as his not-acting has not done what needs doing, the killing of the King. The truth about Claudius does, however, get said at last by "The Mousetrap," which when it springs confirms the Ghost's story and frees Hamlet to act. But to act is to enter the world of Claudius, the corruption of which Hamlet is unable to escape either by not acting in the Prayer Scene or by acting in the Closet Scene.

If corruption attends action and non-action in Elsinore, we should expect it also to attend acts of speech. To some extent it always has, even when Hamlet sought asylum within the non-speech of his wordplay. For wordplay will not content itself with a purely dissociative role, divorcing evil from good. In a world where there are no more marriages, words turn promiscuous, incestuous, and yet repulsively fecund. Once Hamlet has quibbled on "sun/son," for instance, the meanings of that quibble multiply maggotlike even where they are not wanted, as in Hamlet's later remarks to Polonius:

> For if the sun breeds maggots in a dead dog, being a
> good kissing carrion—Have you a daughter?
> *Pol.* I have, my lord.
> *Ham.* Let her not walk in the sun. Conception is a blessing,
> but not as your daughter may conceive; friend, look to't.
> (2.2.181–186)

Hamlet's scurrilities here may be owing, as J. Dover Wilson proposed, to his having overheard Polonius saying earlier, "I'll loose my daughter to him" (2.2.162).[23] If so, then Hamlet's attack here is not aimed at Ophelia herself but at the conception of her implied by her "fishmonger/fleshmonger" father's willingness to use her as sexual bait in the game of court intrigue. Nevertheless, as in Hamlet's double negative advice to Gertrude, the imagistic effect is a portrait of the fair Ophelia as a polluted carcass in which a sun/son may engender worms. And the most likely "son" in question is Hamlet himself, who illustrates in the very process of his wordplay the loathsome kind of "conception" he describes, inasmuch as his puns and metaphors engender through negation ("Let her not walk . . . but not as your daughter may conceive") the very repellence they would obviate.

Procreation conceived of so disgustingly, we can understand why Hamlet should later cry to Ophelia herself, "Get thee to a nunnery; why wouldst thou be a breeder of sinners?" (3.1.122–123). In this scene wordplay again betrays Hamlet by refusing to abide by the categorical neatness of "To be *or* not to be." The key is "nunnery," and I agree with Harold Jenkins, among others, that Hamlet intends that term in its obvious sense: he is seeking to perpetuate Ophelia's virginity by ordering her to a convent, not accusing her of wantonness by playing on the secondary slang sense of "brothel."[24] Unfortunately, wordplay is itself wanton. Each of Hamlet's five repetitions of the word "nunnery" in this scene opens the convent doors a little wider to let the bawdy meanings in, to condemn Ophelia to the whorehouse of the world, and to make of Hamlet the "fleshmonger" he labeled Polonius.

"Nunnery" is an especially promiscuous piece of wordplay because its two meanings are bound together not only by pun but also by metaphor: a convent is a brothel. Erasure is virtually impossible in such a case. But even Hamlet's earlier homophonic pun on "sun/son," in which no metaphoric connection is discernible, resists erasure and dissociation. Given the triggering word "clouds" in Claudius' question ("How is it that the clouds still hang on you?"), the overt meaning of "sun/son" (which in the theater is merely one sound with two possible words in it) is solar. When translated into the insincerities of the court, Hamlet's line says something like "Not so, my lord; I am too much in the sunshine of your royal favor." It is this solar meaning that Hamlet wishes simultaneously to create in the interests of decorum and to erase in the interests of truth. The erasure will make a space for such meanings as:

> Not so, my lord; my recent conversion from your "nephew" to your "son" is by no means welcome.
>
> Not so, my lord; this prolongation of the role of child and "son," even after the death of my father, is unnatural and resented, especially since as "son" of a dead king I should be on the throne and you *my* dependent.
>
> Not so, my lord; in my dispossessed state I am too much out in the sun (or as we would now say, "out in the cold").
>
> Not so, my lord; my grief and discontent, far from being concealed by clouds, are most openly apparent.
>
> Not so, my lord; I am too much subject to your royal scrutiny.

These and other covert meanings express the truth of Hamlet's feelings, and yet they cannot be reached without passing through the cosmetic hypocrisy of his overt meaning ("sunshine of your royal favor"). Moreover, there is no indication that Hamlet's courtly audience ever passes through the surface of his wordplay to register its underlying truths. Thus the verbal situation parallels Hamlet's plight in "showing" his grief for his dead father. That grief can be expressed only in "actions that [he acknowledges] a man might play" within a theater-of-court where false semblance makes all actions suspect (1.2.84). Words, no less than the apparel of mourning—no less than ceremonies, institutions, and customs—participate in the social order. When that order meets with base infection, as in Elsinore under Claudius, words contract the public disease, turn cancerous, and multiply cells of foul meaning at the very heart of truth. The verbal plague in Denmark leaves no one unmarked, perhaps least of all Hamlet, despite his efforts to quarantine himself within wordplay.

One way to demonstrate this would be to regard Hamlet himself as a bit of wordplay, as a pun writ large, for instance. Such a notion has some warrant in the ambiguous reference of the name "Hamlet" to both father and son, an ambiguity that Hamlet seeks to eliminate during the middle of the play by individualizing his name, by refusing to be subsumed under his father's identity. But Hamlet has more than one "father" from whom he disengages himself, not to mention his mother, Ophelia, Polonius, Rosencrantz and Guildenstern, and the whole world of Denmark. His withdrawal behind the pales of self is a retreat into silence and the half-silence of his wordplay, into inaction and the half-actions of his antic disposition. When at last he does speak forth, he goes forth as well, emerging from isolation to insert himself fully into the world, to become related, even to acquire (like a pun) multiple identities: "This is I,/ Hamlet, the Dane!"

This multiplication of Hamlet's identity can be illustrated by two of his major actions, each of which relies upon negation and erasure, and each of which subjects Hamlet to the fate of the dyer's hand. Let us look first at his rewriting of "The Murder of Gonzago." Under normal circumstances the unrewritten play would constitute acceptable entertainment for the court. To the King's "Have you heard the argument? Is there no offence in it?" Hamlet could have replied quite without irony, "No offence in the world" (3.2.232–233). But he should have had to add, "And no truth either. No secrets revealed, no lies exposed." For that reason he does not entertain the court with the inoffensive "Murder of Gonzago." Instead, as in his wordplay, he erases the acceptable meanings of that play and rewrites it as "The Mousetrap," which thus becomes a kind of macro-pun—not an accidental homophonic pun but a deliberately polysemous one whose primary and secondary meanings are too closely related for Claudius to disregard, as indeed he does not: "Give me some light! Away!" (3.2.267).

So much to the good. However, like Shakespeare, whose imagination fused with its own creations during the writing of his plays, Hamlet the amateur playwright was obliged while rewriting "The Murder of Gonzago" to "become" his characters, to speak with the prophetic soul of the Player King, to mouth the hypocrisies of the Player Queen, and to murder with . . . One hesitates here because the identity of the murderer is the one significant departure from the truth that the play depicts. The Player King and Queen are identifiable by their royal crowns and gowns, no doubt, but the identity of the murderer is unavailable to the audience until Hamlet, in a final authorial flourish as he observes the play, announces "This is one Lucianus, nephew to the king" (3.2.242). Nephew to the king, not his brother! As a result the player-murderer is an ambiguous combination of Hamlet, nephew to the present king, and Claudius, killer of the former king; and the theatrical murder tells two truths, one about the past, the other about the future. By substituting "nephew" for "brother," Hamlet makes his own future murder of Claudius issue causally from Claudius' murder (both real and theatrical) of Hamlet's father. But in the process he has had to taint his imagination by participating in the murder of his father as Player King. To kill Claudius *for* his father, in short, Hamlet must (in the Inner-Play) *be* Claudius and kill his father.[25]

This notion of a polysemous Hamlet who, in order to act for one father, must assume the identity of the other is suggested again in Hamlet's story of how during his sea-voyage he stole the King's commission and rewrote it:

I sat me down,
Devised a new commission, wrote it fair.
I once did hold it, as our statists do,
A baseness to write fair, and laboured much
How to forget that learning, but, sir, now
It did me yeoman's service. Wilt thou know
The effect of what I wrote?
Hor. Ay, good my lord.
Ham. An earnest conjuration from the King,
As England was his faithful tributary,
As love between them as the palm might flourish,
As peace should still her wheaten garland wear
And stand a comma 'tween their amities,
And many such-like *as*-es of great charge,
That, on the view and know of these contents,
Without debatement further, more or less,
He should those bearers put to sudden death,
Not shriving time allowed.
Hor. How was this sealed?
Ham. Why, even in that was heaven ordinant.
I had my father's signet in my purse,
Which was the model of that Danish seal;
Folded the writ up in the form of the other,
Subscribed it, gave't the impression, placed it safely,
The changeling never known. (5.2.31–53)

Hamlet's revising of the King's commission duplicates his efforts throughout the play to erase the language of Claudius and substitute his own. Though he is not now taking shelter behind his antic disposition, the stakes in this verbal contest are, even more obviously than before, his life. As in his earlier wordplay, Hamlet must now make his way through, even by means of, the received forms. He may violate the calligraphic standards of the statists by writing "fair," but as the "*as*-es" of his narrative indicate he adopts with something of a vengeance their rhetorical style.[26] If the revised commission now contains Hamlet's meanings within Claudius' words, and if that implies a certain tainting of mind and word, still it appears necessary for the saving of the hero's life. Moreover, the entire stratagem receives a kind of symbolic imprimatur through the serendipitous sealing of the commission with the signet of Hamlet's father. The truant prince assumes the authority of his royal father and for the first time in the play acts in the role of king.[27]

But we need to ask, which father? and which king? For in rewriting the commission Hamlet assumes the authority of his step-father also. Not merely that. A more momentous symbolic act is muted in Hamlet's narrative by his employment of the legal term "subscribed." For it is a peculiar feature of the revised commission that it requires two "signatures," that of Claudius first, then that of Old Hamlet as represented by the impress of his "signet." And Hamlet alone can "subscribe" for both kings.

Much earlier, in the Council Scene, Claudius invited Hamlet to "Be as ourself in Demark." Later, this invitation was metaphorically repeated by the Ghost, who in assigning the role of revenger to Hamlet said in effect, "Be as thy father in Denmark." To be as his father, however, Hamlet seems bound by the illogic of "To be *and* not to be" to be as his stepfather also. Thus the paronomastic Prince, he who can buy up many meanings for the price of a single word, himself takes on the aspect of a quibble as he assimilates, for evil as well as for good, the royal identities of his two fathers. By this time he is indeed "too much in the sun."

TWENTY-ONE

THE LIFE OF DEATH

I HAVE been presenting Hamlet as if he were himself an illustration of the wordplay he engages in, coming increasingly to embody meanings and values from which he had sought to dissociate himself. Despite efforts at erasure and nullification in the play, meanings, values, and images seem indelible, rising into being in the very process of their rejection. If the negative is paradoxically creative in these minor ways, what about the Great Negative itself? In a sense Death underlies all the smaller negatives to which Hamlet and Shakespeare have had recourse throughout the play. These negative presentations of evil and decay are Death's harbingers, forms of death-in-life that usher the play toward its Graveyard Scene, where so many issues come to rest.

One of these issues, mentioned earlier, is the blurring of distinctions. I have suggested that a major object of the hero's quest in *Hamlet* is self-definition, an affirmation of personal identity. However, any such affirmation is bound to be elusive in a Denmark where the boundaries between naturally discrete things dissolve and opposites fuse, where uncles become fathers, mothers aunts, and funerals marriages, where night and day are indistinguishable and the weekdays merge with the Sundays. This demise of distinctions arrives with the kingship of Claudius, who brings Death to Denmark in various forms. That first fatal poisoning in the orchard trails behind it the deaths of Polonius, Ophelia, Rosencrantz and Guildenstern, Laertes, Gertrude, and Claudius and Hamlet themselves. Within that cortege are also the death of Gertrude's love for Hamlet's father, the death of Hamlet's love for Ophelia, the death of reason in Ophelia's mind, and for Hamlet the death of meaning and purpose in the world. Not the least of these is the death of distinctions—for the Great Negative is also the Great Undifferentiator.

Amid all of this Hamlet seeks to maintain differences, to mark off funerals from marriages and uncles from fathers, to draw a line between "To Be" and "Not To Be," and most of all to distinguish himself by means of negation from the court, the King, his own past self, even his role of revenger. Throughout the play Hamlet refuses to "be" in the world of Denmark—although (and this is a major part of his anguish) he remains in and part of that world whether he wants to or not. But then after his sea-voyage he returns and in the Graveyard Scene proclaims for the first time, not what he is *not,* but what he *is:* "This is I,/ Hamlet, the Dane!" This radical reversal of style deserves a close look. Where it takes place deserves, perhaps, an even closer look.

Hamlet's proclamation is set against a background of death—against, that is, a background of undifferentiation and negation. In the Council Scene much earlier Gertrude referred to the "commonness" of the process by which "all that lives must die,/ Passing through nature to eternity," and Hamlet, without having yet earned full title to his irony, replied, "Ay, madam, it is common" (1.2.72–74). In the Graveyard Scene Hamlet discovers just how "common" that process is, in the double sense of "universal" and "vulgar." The grave is notoriously lacking in respect for persons; its invitations are extended to Ophelia and Alexander the Great alike, to Yorick the jester as readily as to Lord Such-a-one. And the Gravemaker himself is no respecter of distinctions. He regards all men as mere bodies returning to dust, he sings unconcernedly at his work because as Horatio observes "custom hath made it in him a property of easiness," and he dismisses the proprieties of Degree ("There is no ancient gentlemen but gardeners, ditchers, and gravemakers") with a presumptuousness that seems to Hamlet symptomatic of the age itself:

> By the Lord, Horatio, this three years I have taken note of it: the age is grown so picked that the toe of the peasant comes so near the heels of the courtier, he galls his kibe. (5.1.138–141)

This general erosion of distinctions is fittingly registered by Hamlet at the edge of a grave that is prepared to house without discrimination its next tenant, as soon as what remains of its former lodgers has been evicted. But Death's commonness, its indifference to distinctions, has not yet made its mark on the Prince. Thus, early in the scene, Hamlet, who had once sought to divide life's To Be from death's Not To Be, attributes to the grave a similar fastidiousness: " 'Tis for the dead, not for the quick," he says, correcting the Gravemaker (who has just said the grave belongs to him), "therefore thou liest" (5.1.126–127). But just as the indeterminateness of death in Hamlet's famous soliloquy blurred his distinctions between To Be and Not To Be, so in this scene the consuming hospitality of the grave effaces his distinction between the quick and the dead. Before the scene is over Hamlet will have leaped "quick" into the grave of the "dead" Ophelia and given the lie to himself by challenging Laertes, "Be buried quick with her, and so will I" (5.1.279).

If Death is "common" in its universal indifference, it is also common in its being vulgar, ordinary, quite literally down-to-earth, and thus physically present. Being physically present is not what one would expect of the Great Negative. It is certainly not what Hamlet expects. Yet Hamlet must come to terms with the material presence of Death before he can come to terms with himself and his own death-dealing mission. That is, throughout the play Hamlet's inability To Be in Denmark, to assume a positive identity, seems bound up with his inability to grasp the Being of Death. The negativity of Death, its indefinable absence of life, is correlative to the negativity of Hamlet, who both asserts and laments his own indefinable absence of identity—his "not being" the emotive Player, the stoic Horatio, the "son" of Claudius, the lover of Ophelia, the militant Fortinbras, or the revenger of his father. Both Hamlet and Death, it seems, have "that within which passeth show," an invisible center.

Thus we should expect Hamlet's "that within" to take a positive turn as Death assumes a more concrete and definable form. Such a turning begins during the sea-voyage when, as he looks upon the royal commission, Hamlet encounters Death imaginatively, not as an "undiscovered country," but as England, where

> not to stay the grinding of the axe,
> My head should be struck off. (5.2.24–25)

This personal encounter with Death proves life-saving for Hamlet, and perhaps soul-saving as well, since his fortuitous escape persuades him that "There's a divinity that shapes our ends" through the agency of a "special providence." But of that, more later.

Even so, an imagining of Death, however vivid, remains within the province of the negative, since by its nature the imagination addresses itself only to things that are not present. In the graveyard, on the other hand, Death is most corporeally present. It appears to Hamlet as neither an undiscovered country nor an imagined axe but an immediately perceptible grave—Death's long house occupied by decaying bodies, worms and dust, and a skull of which he can ask incredulously "This?" and be assured "E'en that." Here he finds the Gravemaker, for whom death is not the Great Negative, object of romantic imaginings, but an indifferent daily presence, an occasion for digging. The grave he now digs, he tells Hamlet, is for no man:

> *Ham.* What woman, then?
> *1 Clo.* For none, neither.
> *Ham.* Who is to be buried in it?
> *1 Clo.* One that was a woman, sir; but, rest her
> soul, she's dead. (5.1.132–136)

The Gravemaker's business is not with what "was"—not with a woman or a man or a "who"—but with what most plainly "is"—dead bodies. For him the graveyard is not where one's life ends so much as where one's work begins. So he builds away at his doomsday houses, drinks his stoups of liquor from Yaughan's, sings his songs, reckons how long pocky corses will last you in the damp ground and, when they have not lasted well, evicts their skulls to make room for newer tenants. He and Death are old familiars these thirty years of Hamlet's life.

But since the "hand of little employment hath the daintier sense," Hamlet finds the Gravemaker's occupation with death more repellent than his own preoccupation with it. Indeed, in a reversal of his earlier style, Hamlet now seeks to negate the Great Negative. In the past Hamlet conceived of Death and suicide as the means of erasing the calamities and evils of life, an ultimate escape from life's "commonness." Now, in the graveyard, he conceives of life as an imaginative escape from the commonness of Death. That is, by restoring to the nameless skulls their imagined identities as Lord Such-a-one, the politician, the courtier, the lawyer, and of course Yorick, Hamlet attempts to erase the rude fact of death. But his attempts fail, in two stages. First, the free play of his imagination is placed under strict arrest by the Gravemaker's matter-of-fact identification of one skull as belonging not just to anyone but to a very specific someone: "This same skull, sir, was Yorick's skull, the King's jester." All Hamlet's speculations dwindle into the starkness of his "This?" and to the silence that follows. But then, recovering somewhat, he resumes his efforts, only to find that something fails him:

> Alas, poor Yorick! I knew him, Horatio, a fellow of infinite jest, of most excellent fancy. He hath borne me on his back a thousand times. *And now how abhorred in my imagination it is!* My gorge rises at it. Here hung those lips that I have kissed I know not how oft. Where be your gibes now, your gambols, your songs, your flashes of merriment that were wont to set the table on a roar? Not one now, to mock your own grinning? Quite chop-fallen? . . .

Hamlet's attempts to negate Death by painting over it images of former life are defeated by the implacable fact of the skull in his hand, which remains all too visible beneath the remembered face Hamlet superimposes over it. Life so conceived of, or reconceived of, turns "abhorred in [his] imagination" and dies a second death there, leaving as the residue of its merriment an imperturbable bony grin. Thus Hamlet's concluding advice to "my lady" about the elemental color beneath all her paintings—

> Now get you to my lady's chamber, and tell her, let her paint an inch thick, to this favour she must come. Make her laugh at that.

—seems to acknowledge the futility of his own verbal cosmetics in this scene, the inch-thick images of life with which he has painted Death's skull.

At this point Hamlet seems to accept Death as a literal presence that cannot be negated. His next remarks are in keeping with such a view. In them his "imagination [traces] the noble dust of Alexander" and Caesar as it makes its way through earth and loam toward beer barrels and flawed walls. It seems noteworthy that Hamlet imagines these antique heroes not as they were in life, as he has just done with the various skulls, but as they might become in death. His tracing of this dusty degenerative process is hardly reassuring, and yet in it Death is not wholly negative, an absence of life, but has a vitality of its own as it performs its metamorphoses within the earth. Death's subterranean doings are a phase of that ceaseless revolution of life in which, as Hamlet tells Claudius elsewhere, "a king may go a progress through the guts of a beggar" (4.3.31–32). Thus Death is possessed of a macabre form of life, albeit the life of decay.

Even the grave shares something of this life. In its purely negative aspect the grave has its grisly import as the dread wormy hole into which Hamlet and Laertes descend in unwitting rehearsal of a descent tomorrow as sadly terminal for them as that of Ophelia today. But the grave is not just an absence of earth and life. In it Hamlet and Laertes compete with one another in grief for the lost Ophelia, and the quick and the dead compete for rights of occupancy. For a few moments the grave is procreant with life and conflict, a place from which the two men issue as well as enter, and from which they issue with new or renewed purpose: Laertes obsessed with thoughts of a revenge to be pursued whatever the circumstance, and Hamlet conscious of how in the teleology of Providence circumstances themselves will cause what-is-to-be to be—

> Let Hercules himself do what he may,
> The cat will mew and dog will have his day. (5.1.291–292)

Or, since our theme has been wordplay, one might suggest that in its sudden bodily plenitude the vitality of this grave is like the "life" of a pun, within whose verbal confines different and sometimes conflicting meanings compete for semantic space. And as a pun marries in one sound its several meanings, so the grave marries Hamlet and Ophelia despite their mortal division, and even embraces Hamlet and Laertes despite their deadly strife. Long ago Hamlet directed Ophelia to a nunnery with the words, "I say, we will have no more marriage" (3.1.149–150). In the broad sense of "community," marriage has indeed vanished from Denmark, abolished not by Hamlet's "I say," however, but by the "unsayings" of Claudius, the unspoken truths masked in lies that have contaminated such marrying values as love, honor, and faith. And yet in the absence of these more elegant conjunctions Hamlet discovers with a kind of corporeal force in this scene that all men are bound *by* what they are ultimately bound for. "Ay, madam," he could say now, less to his mother perhaps than to My Lady Worm, "it is common."

TWENTY-TWO

Beyond Negation

TO SUM up from the standpoint of negation: Hamlet's response to the disclosures of the Ghost is to erase from his world its fairer surfaces, which he now regards as mere cosmetics painted over the moral and mortal ugliness of life. In his disillusionment he would address a terminal "No" to life itself, were it not that the Everlasting had pronounced against suicide a prior "No" and that the Great Negative might prove, not an escape from, but an intensification of life's evils. Denied the nihility of death, yet unable to affirm life, he lives on as an outsider-within, a detached and embittered ironist commenting upon an abruptly fallen world. His speech is parenthesized in wordplay and his actions in madplay, a bracketing whereby he seeks to remain free of the contaminations of Claudius' court. But the freedom he achieves is a half-freedom, a freedom not "To Be" but only "Not To Be," the freedom to say with Iago, "I am not what I am." In the graveyard, however, Death and Hamlet both acquire a positive presence—the irreducible "This" of Yorick's skull corresponding to his own "This is I." For Hamlet fully "To Be," it seems, he must experience in the graveyard, under the tutelage of the Gravemaker, what it is "Not To Be." For his own identity to crystallize, he must come to the place where all identities dissolve.

Hamlet's "This is I" is addressed of course to Laertes, Claudius, and the other mourners—a revelation of his identity as he probably throws back his sea-scarf to show himself. However, it also seems a response to the scene itself. That is, given the profound symbolic suggestiveness of the graveyard episode, Hamlet's "This is I" assumes the nature of a mythic challenge, somewhat like that of Childe Roland when he arrives at last at the Dark Tower in Browning's enigmatic poem. If so, Hamlet's cry is directed to the Gravemaker and his skulls and songs, to the dust of buried heroes, to the remorseless process by which sweet maids wither into corpses, meanings into absurdity, and purpose into indifference. It is a gesture asserting the primacy of the individual in a wasting world, the tragic self at grave's edge.[28]

But alas, at grave's edge indeed! For the histrionic force of Hamlet's gesture inevitably submits to the measure of its mortal context. It possesses a good deal of the self-consuming irony of his later "proof" of the magnitude of his love for Ophelia—his death-defying, death-inviting "Be buried quick with her, and so will I." In scarcely more than an hour Hamlet's proposed death pact with Laertes will have been consummated, and not long thereafter the Gravemaker will have resumed his labors in clay—the sounds of his spade and his songs distantly counterpointing the earnest voice of Horatio telling Hamlet's story to the yet unknowing world.

It is not unusual for a sense of illumination to invade the tragic hero as his fate closes around him, but what he sees is normally his own diminished self and a disintegrating world that, except for him, might have remained intact. "Hell," as Thomas Hobbes said, "is truth seen too late," and nowhere is this better illustrated than in the famous speeches by Macbeth at the end of his play, "My way of life/ Is fallen into the sere, the yellow leaf" and "Tomorrow and tomorrow and tomorrow." But in contrast to Macbeth, Hamlet moves in his play not toward meaninglessness but toward meaning. By a route we cannot fully trace, but which leads significantly through the graveyard, Hamlet passes from the void of "And yet, to me, what is this quintessence of dust?" to the conviction that "There's a divinity that shapes our ends,/ Rough-hew them how we will," and from the febrile irresolution of "To be or not to be" to the calm recipience of "Let be." Though by this point he is by no means untainted himself, Hamlet does not, like Macbeth, regard himself as having shriveled with evil. Instead, he has graduated from a desire for self-dissolution—"O, that this too too solid flesh would melt"—to the almost triumphant self-assertiveness of "This is I,/ Hamlet, the Dane!"

After all his negations, then—his ironic not-meanings, his disclosures of a not-self, his advertisements of his not-revenge—Hamlet has emerged, like a developed photograph from its "negative," into clarity and resolution. This movement beyond negation is prerequisite to Hamlet's revenge. For revenge is an act of destruction, and negation destroys nothing. Far from it: by its presentation of the absent, its affirmation of the denied, it not only does not destroy but actually creates what it nullifies. The evils to which Hamlet has repeatedly said "Not this" have fed and thrived on his disavowals, multiplying cancerously within the imagistic body of the play. What is "rotten in Denmark" cannot be negated; it can only be destroyed by an act of revenge against Claudius.

At this point Hamlet and Shakespeare are in like situations, each with debts falling due. To fulfill his obligations to his father, Hamlet must kill the King. Similarly, to fulfill *his* obligations to the genetic and generic father of *Hamlet*—the *Ur-Hamlet* of Thomas Kyd—Shakespeare must transform his hero from self-searching truant into active revenger. But this transformation is a matter of considerable delicacy. As revenger, Hamlet must be capable of the act required of him, and yet, as a character in whom Shakespeare and his audience have a profound investment, he must not lose dramatic stature—as he might, for instance, if he charged pell-mell toward regicide like Laertes on his return from France or, alternatively, if he sidled toward his revenge with guile and subterfuge like the smiling King himself.[29] Moreover, if the act of revenge requires an actor suitable to perform it (and hence a transformation of Hamlet), it also requires a context or scene suitable for its performance (and hence a transformation of Denmark). The act, the actor, and the scene of action interpenetrate with considerable complexity to fashion the end of the play. Without losing sight of the negative, let me glance at what Kenneth Burke calls the "act-agent ratio" (the transformation of Hamlet) and then at the "scene-act ratio" (the transformation of Denmark) as these bear on the ending.[30]

Whereas earlier we stressed the positive aspects of negation, now we would note the negative aspects of the positive, particularly the toll taken upon Hamlet by the act he is to perform. For we gain the new Hamlet, the Hamlet bent like a bow for revenge, at the high cost of losing the multifaceted Hamlet engrossed in the business of not-revenge. As I argued in the previous chapter, action defines the actor by abstracting a universal identity from the fullness of his concrete particularity—in the present case, by refining Hamlet the complex individual into that generic entity, Hamlet the revenger. The process is very like that by which the truant but colorful Prince Hal yields at Westminster to his actional self, the militant King Henry the Fifth, or like that by which the self-singing, ineffectual Richard II yields at Flint Castle to the taciturn, pragmatic Bolingbroke.[31] As Hal, versatile prince of holidays, advances toward his other and lesser self, the stringent King Harry, so the various-minded Danish prince makes his digressive way toward a symbolic union with his Norwegian counterpart and displaced lesser self, Fortinbras, soon to be King of Denmark in his stead.

The loss of concrete "being" in the conversion of Hamlet to revenger is particularly costly because Hamlet's being is founded on not-being, a more capacious realm. As long as Hamlet asserts not what he is but only what he is not, his undisclosed real self takes on richness and diversity *in potentia*, as analogously all forms are contained in a blank canvas, and all sounds in silence. Moreover, Hamlet's richness of individual being profits from the identities he disclaims. When he denies being the son of Claudius, the lover of Ophelia, the passionate Player, the dispassionate Horatio, a reincarnate Nero, or the honor-struck Fortinbras, his denials lend possibility to what is denied. It is not, we see, that Hamlet is none of these, but that he is none of them entirely, for he has characteristics of them all. They exist as tangents to the undescribed curve of his concealed self, as visible traces of "that within which passeth show." During the period of his delay he can momentarily try on these optional selves without committing himself to any of them, because he is safeguarded by the "If I were" of the negative imagination, with its tacit coda, "But of course I am not." Were he to identify wholly with any of these "parts" Hamlet would contract into a portion of himself, becoming a synecdoche of his total self, and this he refuses to do. Until the Duel Scene, that is, when the play presents him with a part he cannot refuse.

Thus the act-agent ratio features reduction, loss, and sacrifice as well as achievement. As the act of revenge will take form by negating other acts that might have been, so the revenger, as a product of that act, will acquire his identity by erasing the selves he might have been. Even the seemingly positive shape of the play itself evolves in this negative fashion. The essence of the tragic entelechy, as Aristotle emphasized, is an erasure of options—a reduction of the plot from the possible to the probable, from the probable to the inevitable—so that the final form of the play consists not only in its being what it is but also in its not being what it might have been. Thus Gertrude's parting remark to Ophelia: "I thought thy bride-bed to have decked, sweet maid,/ And not have strewed thy grave" (5.1.245–246). The isolation of the tragic hero, his death, the random deaths of others, the pervasive sense of loss within the community—these aspects of the fictional world of Thebes, or Elsinore, are reflected also in the contracting form of the play itself as it diminishes toward its own end.[32]

Yet the end must come, the awaited act be performed, and the hero readied for that act at whatever cost. But if the act calls for an agent, a revenger, it also calls for a scene, and thus far the scene of Elsinore has been deterrent rather than conducive to the act. Instead of construing the Ghost's command in its most limited sense, as meaning simply "Kill the King," Hamlet assumes that the very "times" themselves are "out of joint" and that he is cursed with the task of setting them right. Thus he transforms a local issue into a cosmic dilemma. Of course the times *are* out of joint, most painfully, but the extent of their dislocation is magnified in the imagination of their self-appointed physician. To Hamlet, nothing is free of affliction, least of all himself. Sensing contamination about and within him, he labors less to act than to explore in morbid fascination his scene of action in its broadest dimensions, as an entire world given over to evil, disease, degeneration, false seeming, and death. The impossible task of coming to terms with such a world, with life's most baffling mysteries and paradoxes, seems prerequisite to his revenge.[33]

If in the final hour before the duel Hamlet has come to terms with his scene of action it is because he has envisaged a scene-atop-a-scene, placing the vexations of earthly "times" within the overarching frame of Providence. His rash theft of the commission at sea was, he now realizes, endorsed by that "divinity that shapes our ends"; his fortunate possession of his father's signet ring—"Why, even in that was heaven ordinant"; and as for his own possible death—"There's a special providence in the fall of a sparrow." Even here, in the invocation of a transcendent order, the negative plays its sly role. For if earthly affairs must await their final definition in terms of the mysterious criteria of divinity and providence, then life as Hamlet has sought to know it is in some measure nullified and his own judgments called in question. Things down here look considerably different from the perspective of the eighth sphere, as Chaucer's Troilus assures us. Not of course that Shakespeare suddenly reduces life in Elsinore to a mere episode in a Divine Comedy. But he does, in a fairly unobtrusive manner, call upon a suggestion of life as seen through a glass darkly. Thus Hamlet, having broken all the mirrors of false-seeming hung by Claudius in the castle at Elsinore and encountered in their absence a world shot through with vileness and evil, now comes upon a subtler mirror still, one not hung by Claudius but fashioned by the very nature of the human situation in a fallen world. The "reality" he thought he had discovered is, it seems, but another illusion, sufficiently indeterminate at least to cause him to abandon his "irritable reaching after fact and reason," in Keats's words, and to embrace instead something like the principle of Negative Capability, "Let be." If Hamlet experiences a tragic *anagnorisis*, it is not the recognition of an Aristotelean *hamartia*—"the stamp of one defect" or a "particular fault" from which he takes corruption (1.4.23 ff.)—but rather a more general acknowledgement of human finitude. He has traced with the finger of imagination the shadowy "not" that marks the boundaries of what man can know and do.

Hamlet's "Let be" lies somewhere between the implied "Yes" he accorded to the world in those distant days when he was "the expectancy and rose of the fair state" and the explicit "No" he has directed at life since the death of his father. Similarly, his "readiness," as it extends to his revenge, lies between the extremes of inaction (including his vicious "not-killing" of Claudius at his prayers) and rash violence (his murder of Polonius). It is a species of pre-action, or, more precisely, a species of pre-*re*action, for Hamlet is tensed less to act than to react, not like a sprinter at the blocks knowing exactly what he is going to do when the gun goes off, but like a fencer who does not know when or where his opponent will make his thrust.

As I said before, Shakespeare is anxious that Hamlet not lose the sympathies of his audience. To that end he has displaced the ultimate responsibility for action from his hero to the scene outside him, which comprises the immediate plot of Claudius and the enveloping "plot" of Providence. This strategy of situating motives for action in the scene rather than in the agent is what the Gravemaker would call *se offendendo* (actually *se defendendo*), which is carefully defined in his discussion of guilt and innocence in death by drowning:

> Give me leave. Here lies the water; good. Here stands the man; good. If the man go to this water, and drown himself, it is, will he, nill he, he goes, mark you that. But if the water come to him and drown him, he drowns not himself. Argal, he that is not guilty of his own death shortens not his own life.
>
> (5.1.15–20)

As the water seeks out its innocent victim, so the murder of Claudius seeks out Hamlet. That murder is instinct within Claudius' plot, within the duel-as-scene, and Hamlet quite literally obliges the duel to come to him when he tells Osric, "Sir, I will walk here in the hall. If it please his Majesty, it is the breathing time of day with me" (5.2.171–172). Thus, as emblematic of this protective strategy, we do not see Hamlet enter to the King and court but rather the King, the Queen, and "all the State" enter to Hamlet. Argal, as the Gravemaker would say, he that does not initiate his revenge is not wholly guilty of its performance.

The revenge itself is Hamlet's last act of erasure, a final "Not this" directed at Claudius, the source of rottenness in Denmark—"this canker of our nature" that festers so near to the heart of the state that only death can cure it (5.2.69). However, the killing of Claudius is less central to our present theme than the killing of Hamlet. Not the actual killing, the death of the body, but the death of what Hamlet has been. As we have seen, at the end of the play various forces act to reduce Hamlet the complex individual to a generic identity, to a part within a larger whole—a member of Death's entourage, a player in the drama of Providence, an agent for his father. Most important, though, is the act of revenge itself, which Hamlet will bring into being but which will reciprocate by bringing Hamlet into being in its image.

And that image is hardly a pleasing one.[34] For the revenge is no swift, clean, classical stroke imparting a certain grace and beauty even to murder. Perhaps the aesthetics of murder paled for Shakespeare in the writing of *Julius Caesar*, where the ritual sacrifice intended by Brutus turns into the bloody hacking of Caesar's body.[35] In any event, although death is the necessary culmination of dramatic form in the Globe theater, in the palace at Elsinore it represents the dissolution of form when the fencing match, the stylization of violence, explodes into havoc, leaving the stage a heap of bodies. "The sight is dismal," the Ambassador says, and even the hardened Fortinbras observe that this sight "Becomes the field, but here shows much amiss" (369, 404). And in the faces of those

> that look pale and tremble at this chance,
> That are but mutes or audience to this act,

Hamlet sees the truth of the Ghost's remark long ago when he called "Murder most foul, as in the best it is" (1.5.28). The "taint" the Ghost warned him against, the poisonous contamination he has sought to avoid, is now upon him; it has waited within the very nature of the act he was bound to perform. The most "potent poison" now is not that which curds the "thin and wholesome blood," as it did his father's, but that which corrodes the sense of self, so that Hamlet suddenly feels as debased as Macbeth:

> O God, Horatio, what a wounded name,
> Things standing thus unknown, shall I leave behind me!

The name whose integrity he wants to restore is no longer his father's but his own, hard won in private strife. In calling on Horatio to tell his "story" Hamlet seeks release from the bondage of his act and the degraded identity it has conferred upon him—a retrieval of what he has "been" from the consuming force of what he has "done."

Thus Hamlet's anxiety about his "name" at this point—at a time when the radical "I" seems to have been lost in the generic nature of action—is analogous to his defiant avowal "This is I" in the graveyard where all identity is lost in the wasting action of Death. With Death now at hand, "strict in his arrest," and the Gravemaker waiting, as always, Hamlet's body is about to join Alexander and Yorick and Ophelia—and ultimately, under the ministrations of My Lady Worm, to become one with the indifferent clay. But Hamlet himself, if not his body, can escape the nothingness of Death and, in the story Horatio tells, achieve a continuing verbal life, a "name." In keeping with the negative mode he has pursued for so long, he will acquire a kind of life even in death. It seems fitting that at the end Hamlet is both "to be and not to be."

PART III

Ends and Means: Artistic Arrest and Functional Erasure

TWENTY-THREE

MIDDLEMEN

SHAKESPEARE'S failure to supply Hamlet with a name that would distinguish him from the king who begot him or from the play that begot both him and the king seems an instance of verbal stinginess so uncharacteristic of the prodigal playwright as to beg for interpretation. As a result, critics are led almost inevitably to the names of Welsh island towns like Llanfairpwllgwyngyllgogerychwyrndrobwillllandtysiliogogogoch, into whose polysyllabic depths most literary mysteries disappear. Unfortunately, however, *Hamlet* has an abundance of mysteries. If one of them consists of the absence of a name, another arises from the unexpected presence of a name.

In Act 4, Scene 7, the King seems about to placate the revengeful Laertes by telling him of his own plan for eliminating Hamlet in England, when a Messenger arrives with letters from Hamlet himself:

> *King:* From Hamlet! Who brought them?
> *Mess.* Sailors, my lord, they say; I saw them not.
> They were given me by Claudio. He received them
> Of him that brought them. (4.7.38–41)

Maurice Charney's response to this passage probably speaks for all of us: "Who is this cryptic Claudio, whose name is the Italian form of Claudius?"[1] He also acts for all of us, I suspect, in leaving the question unanswered. Yet it is a question to be asked. Here, after all, is a dramatic nobody, a noncharacter who never appears on stage, who merely receives some letters and hands them on (or is reported to do so), and yet who is dignified by a name that is arrestingly similar to that of Denmark's murderous monarch.[2] How to account for it?

Not by direct critical assault, surely. In a play in which directions, as Polonius assures us, come only through indirections, no one but Laertes storms the castle demanding immediate answers. And Claudio would hardly reveal his secrets to the rampaging Laertes; he would shrink further back into the recesses of the play, hiding his identity in his name. Who could induce him to speak? Perhaps Horatio. Horatio has little success inducing the Ghost to speak, but he fares much better with Hamlet, to whom he spends much of his time listening. That in itself is rather odd—that Horatio, whose name is etymologically associated with speaking, should prove such a great listener throughout the play. Indeed, Horatio, if we study him more closely, begins to take on some of Claudio's mysteriousness. Let us, then, take a roundabout route to Claudio by asking first what Horatio can tell us about him.

Some of Horatio's oddnesses have been catalogued by John Draper, who observes for instance that although Horatio is clearly Danish, he seems surprisingly ignorant of Danish affairs:

> Though a "gentleman," as his relations with the Prince and Marcellus show, yet he is strangely ignorant of both the customs and the personalities of the court. The "heavy-headed revel" of Danish royalty must be explained to him. Apparently, he has never heard of Yorick. Hamlet has to tell him who Laertes is. He has not the pleasure of Osric's select acquaintance. . . . Not only is Horatio ignorant of the court; but, even more important, the court, until the last scenes of the play, seems ignorant of him. He does not move in the charmed circles of Rosencrantz and Guildenstern, whose death brings from him only casual remark. The King asks him, but not by name, to follow Ophelia; and, though present in the scene, he is almost entirely ignored. Shakespeare deleted him from the court-scenes until the play-within-the-play, to which Hamlet seems to give him a special invitation. Neither Claudius nor Gertrude thinks to use him to discover Hamlet's malady, though both, for different reasons, make every effort to learn it, and do not hesitate even to use Ophelia. As late as the fourth act, when Horatio and a "Gentleman" enter to tell the Queen of Ophelia's madness, the "Gentleman" does practically all of the talking; and Horatio seems to be shown on stage only to remind the audience that, though Hamlet has started for England, his friend remains at Elsinore. Indeed, he seems to be unknown to the court until Act V, when the King asks him to follow Hamlet and "wait upon him" . . . . A gentleman

> and a scholar, but not of the court or capitol—where then did Horatio belong in the closed corporation of Renaissance society . . . ?[3]

Where indeed? And even more important, where does he belong in the not entirely closed world of Shakespeare's play?

Let me add to this puzzling picture of Horatio a few other disturbing details. It is a bit strange, for instance, that in the opening scene of the play Horatio speaks of having seen Hamlet's father twice, once when he the "ambitious Norway combated"—that is, when he fought old Fortinbras, that being (as the Gravemaker informs us later) "the very day that young Hamlet was born" (5.1.147–148). The strangeness of this will not emerge until the following scene when we learn that Horatio is the contemporary not of old Hamlet, as it first appeared, but of young Hamlet, his Wittenberg schoolfellow. At this point the discrepancy is no great matter, merely a slight friction.

That friction increases somewhat when Hamlet enlists the aid of Horatio in springing "The Mousetrap" on the King, pointing out that "One scene of it comes near the circumstance/ Which I have told thee of my father's death" (3.2.75–76). That Hamlet has confided in Horatio about the nature of his father's death comes as a considerable surprise if we recall his frenetic insistence on secrecy after the departure of the Ghost in 1.5. To Horatio specifically he said,

> It is an honest ghost, that let me tell you.
> For your desire to know what is between us,
> O'ermaster it as you may. (1.5.139–141)

If Hamlet has nothing to say to Horatio here, Horatio takes his revenge during the remainder of the play with his own stoic silence. Thus G. Wilson Knight, observing that Hamlet has no friend except Horatio, adds, "and Horatio, after the Ghost scenes, becomes a queer shadowy character who rarely gets beyond 'E'en so, my lord,' 'My lord—,' and such-like phrases."[4] After the Play Scene Horatio next appears—if we discount some out-of-character lines sometimes attributed to him in 4.5—as the recipient of Hamlet's letter in 4.6. He reads the letter aloud but says virtually nothing in his own person. And then during the Graveyard Scene he is reduced, as Knight says, to "E'en so, my lord" speeches, which remain his stock-in-trade until the death of Hamlet, at which point he himself is dissuaded from suicide by the dying Hamlet, who pleads the need of his narrative voice to tell his "story."

What we have in Horatio, then, is a character who stands out clearly enough during the Ghost scenes, who grows increasingly silent and indistinct as Hamlet's confidant, but who comes again to the fore in the closing moments of the play as the one person who can convey what has happened. He who has been a silent listener, the ear of the play's secrets, will now fulfill the promise of his name by becoming the play's official voice:

So shall you hear
Of carnal, bloody, and unnatural acts,
Of accidental judgments, casual slaughters,
Of deaths put on by cunning and forced cause,
And, in this upshot, purposes mistook
Fallen on the inventors' heads. All this can I
Truly deliver. (5.2.382–388)

TWENTY-FOUR

DISAPPEARING MESSENGERS: CLAUDIO AND HORATIO

HORATIO AS narrative middleman, or rather as midwife who can "truly deliver" the story that was in danger of dying with Hamlet, has something in common with that mysterious middleman with whom we began this discussion. For Claudio is also involved in a process of communication. A less grand process, to be sure—merely a letter from Hamlet to the King, not the entire "story" of *Hamlet*—but one that is surely more complicated, especially as the Messenger describes it. To repeat: when the King asks who brought the letters, the Messenger replies,

> Sailors, my lord, they say; I saw them not.
> They were given me by Claudio. He received them
> Of him that brought them. (4.7.39–41)

Instead of playing Horatio's role in a three-phase process ("Hamlet – Horatio – Danish audience"), Claudio appears, or disappears, amid a plethora of go-betweens, thus:

Hamlet – Sailors – Claudio – Messenger – King.

One messenger, we should have thought, would serve well enough. Two would seem something of a luxury. But three is conspicuously redundant. And there in the exact middle is Claudio, the very soul of redundancy inasmuch as he neither receives the message from Hamlet nor delivers it to the King but merely passes it from one messenger to another. In fact, the harder one looks at Claudio, the harder he is to see. That is because transparency is a characteristic of functionaries, even those who are visible, as Claudio is not. For instance, in this scene we watch the Messenger—whose personal identity is already erased by his generic title—as he enters and crosses the stage toward the King, and for the moment our attention is caught by the appearance of a new character. But once he is identified as "Messenger" his individuality becomes subordinate to his function, and our attention shifts from him to the message that passes by means of him to the King. To invert Macbeth's remark, surmise is smothered in function.

This erasure of the messenger by his function is graphically illustrated by the Ghost, whose first two appearances establish it as a presence of intrinsic albeit cryptic fascination. Ignorant of its purpose, we, like Horatio and the guards, are preoccupied with its mode of being (or nonbeing), with its immediate appearance, with its "Is-ness." When it begins its narration to Hamlet, however, its presence succumbs to its function. The ghostly personage fades into its role as messenger, and that role pales beside the arresting presence of the message itself, with its disclosure of murder. In fact, the Ghost as messenger disappears into its own message. That is, before it reveals its tale-telling function the Ghost distinguishes itself from Hamlet's father—"I am thy father's spirit . . . . If thou didst ever thy dear father love—" (1.5.10, 24)—but as soon as it begins to discharge its function its "I" stands not for the Ghost itself but for Hamlet's father: "Thus was I, sleeping, by a brother's hand/ Of life, of crown, and queen, at once dispatched" (75–76). The Ghost becomes ghostlier still at the end of this scene when not merely is it obliged to disappear from the castle but even its presence there is erased when Hamlet requires abjurations from his friends.

What, then, of Claudio? As the very essence of functional redundancy he, we should think, would come most completely under erasure. And yet he is gifted with a name that seems to lend him an identity apart from his role as mere messenger, that seems to elevate him from functionary to personage. But if this is the case, why does Claudio never put in an appearance on stage, and thus remain an enigma?

Perhaps because an enigma is what Shakespeare wanted Claudio to become. Perhaps because the question "Who is this Claudio?" is what Shakespeare wanted us to ask. And perhaps because the relation of names and functionality is an issue he wanted us to ponder. For if we wonder why Claudio the redundant middleman is given a name, we can hardly help reflecting on the familiar namelessness of functionaries, like that of the manservant who died after years of service to a famous duke and was honored by his employer with a plaque on which his name was misspelled. The perfection of functionality is thus achieved in a territory of namelessness, in the anonymity of machines that perform without a characterizing style; the better they work the more invisible they become. For his part, Claudio is so perfect at his postal work that, had he not been named, we should never have known he existed. Paradoxically, then, Claudio simultaneously appears and disappears in the process of being named.

Still, we might object, Claudio remains a fertile source of obscurity. For one thing, why should Shakespeare make such an issue of names and functions; and for another, why is the name "Claudio" so like "Claudius" as to imply a connection between the two? Well, as for names and functions, that will become clearer as we go along, especially if we cultivate a willing suspension of disbelief. As for the second question, it is well asked, for what can the menial Claudio share with the King of Denmark? If we observe that the only thing Claudio possesses, besides his troublesome name, is his functional role in a communicative process, then we might realize that Claudius also plays a functional role in a communicative process. Of course he is the recipient of Hamlet's letter. But, more than that, he is also its last messenger. For the ultimate recipient of Hamlet's letter is the audience in the Globe theater, and the function of Claudius at this point in the play is to convey the letter to that audience by reading it aloud. Transient though it may be, that function takes a temporary toll on the King. Not that he acquires Claudio's genius for self-erasure, but that for the moment his identity as King of Denmark, as a "real" person in Elsinore, is subordinated to his function of providing information to the audience in the Globe. A slight disturbance, a tremor in the theater, puts the King out of focus for an instant. For the "real" King of Denmark, upon receiving a letter from a man he has sent to his death, would surely withdraw to examine its contents in private, or at least read it silently. He would not announce immediately, as he does, "Laertes, you shall hear them" (4.7.41). For a moment, then, the King is sacrificed to a convention of the theater.

Thus Claudio offers us a small opening onto a large issue: the functional dimension of messages, of communication. Horatio would have led us there also. If we note the momentary functionality of the King in reading Hamlet's letter aloud, we will recall that some forty lines earlier Horatio performed the same function in reading aloud another letter from Hamlet (4.6.13–30). And that will enable us to put a different interpretation on the fact mentioned by John Draper in the passage I quoted earlier, that in Act 4 "Horatio seems to be shown on stage only to remind the audience that, though Hamlet has started for England, his friend remains at Elsinore." Why the audience should need this reminder, Professor Draper does not say and I cannot imagine. But what is apparent when Horatio reads Hamlet's letter aloud is that Shakespeare had to leave him behind in Elsinore to provide Hamlet with communicative access to the audience in the Globe (the King being available only for limited duty in that respect). Horatio is Hamlet's messenger in residence at Elsinore.

There are many messengers in *Hamlet*. To begin with, there is the Ghost conveying messages from the otherworld to Hamlet. Then there are the King's ambassadors carrying messages back and forth between Denmark and both Norway and England; Reynaldo carrying "intelligence" between Paris and Elsinore; the Messenger who brings news of Laertes' riotous arrival from France; the Sailor who brings Hamlet's letter to Horatio; the series of messengers who convey Hamlet's letters to the King and Queen; and both Osric and a Lord who invite Hamlet to duel with Laertes and return his acceptance to the King. As the case of Reynaldo suggests, the role of messenger merges easily into that of spy, so that we can regard the "sifting" Rosencrantz and Guildenstern, and Polonius, and even Ophelia and the Queen, as the King's undercover ambassadors to Hamlet. If we increase our metaphoric tolerance, we see all informers and tellers as messengers: Horatio telling Hamlet of the Ghost, Ophelia recounting Hamlet's antic visit to her chamber, Rosencrantz informing Hamlet about the War of the Theaters, the "tragedians of the city" themselves conveying to the King the incriminating message of "The Mousetrap," Hamlet delivering his message of mordant morality to his mother in the Closet Scene, his mother the Queen giving her lyric account of Ophelia's drowning, Hamlet telling Horatio of the events of his sea-voyage—and all of these leading to the master message, the story of Hamlet to be conveyed to the yet unknowing world by Horatio.

However (to return to Horatio), these messages are all delivered within the world of the play, whereas the messages carried by Horatio are with rare exception delivered to the Globe theater. Perhaps this implies an explanation of the inconsistency and shadowiness that critics have noted in Horatio, for these are products, it seems, of Horatio's being less a realistic personage in Denmark than a function of Shakespearean dramaturgy.

But that is not how he begins. Throughout the first act he is on the whole as realistic and substantial a citizen of Elsinore as Polonius. He is invited to the midnight rendezvous with the Ghost because he is a skeptic, and a friend of Hamlet, and because as a scholar he can address the Ghost in Latin, the official tongue of preternatural diplomacy. He continues to be invested with full fictional reality when he reports the ghostly experience to Hamlet and when he accompanies Hamlet to meet the Ghost at the end of Act 1.

But then Horatio disappears from the play throughout Act 2, and when he returns in Act 3, Scene 2, it is as Hamlet's accomplice in staking out "The Mousetrap." At this point complications arise. Hamlet, we discover, has inexplicably revealed the secrets of the Ghost to Horatio. Why? Because he needs Horatio's assistance in observing Claudius during the play. But surely Hamlet could keep an eye on Claudius himself, as indeed he does. The fact is that Shakespeare needs someone for Hamlet to confide in. Hamlet is going to set "The Mousetrap," but the audience must be informed of his intentions if the scene is to make theatrical sense. Of course Hamlet could supply that information in an aside or by means of soliloquy; but it would require an awkwardly lengthy aside, and Hamlet has already delivered four soliloquies, the last a little over 150 lines ago. Well, then, let Hamlet simply tell someone about it. Let him tell Horatio. But in that case Horatio will have to know about the Ghost's revelations. Then let Hamlet mention having already enlightened his friend on that score. And why Horatio? Because Hamlet can trust him. Let Horatio therefore be made trustworthy. This requires of Hamlet a speech extolling the stoic virtues of Horatio—a speech that seems, however, disproportionately ennobling, causing William Empson for instance to remark somewhat ungenerously, "then, quite unexpectedly, [Hamlet] fawns upon Horatio as a man who is not 'passion's slave,' unlike himself, and we advance upon the Play-within-the-Play."[5] Hamlet's speech, which by virtue of its portrayal of Horatio's character seems to inscribe his friend within the realistic context of Elsinore, paradoxically erases him from that context and situates him instead within the Globe theater, or at least transforms him into a slightly diaphanous character moving between the

Danish palace and the English stage. Thus Hamlet's characterization of Horatio is, like the naming of Claudio, both an affirmation and a denial of his existence in Denmark.

Increasingly, Horatio becomes the recipient of Hamlet's confidences. When Hamlet is sent to sea, Horatio is left behind so that he can read Hamlet's letter aloud, so that he can be informed at length of Hamlet's adventures when he returns, so that he can hear of Hamlet's misgivings before the duel and be told that "the readiness is all." He is placed in the ear of all of Hamlet's conference; and in this role he is both Hamlet's steadfast friend and his constant betrayer, holding his tongue in the castle but telling all in the theater. Like Ophelia in the Nunnery Scene and Gertrude in the Closet Scene, both of whom induce Hamlet to speak so that he may be overheard by the King and Polonius, Horatio provides occasions whereby Hamlet may be overheard by an eavesdropping audience in the Globe, especially in Act 5 where Horatio becomes a microphone for what are, in effect, displaced soliloquies by Hamlet. During the Graveyard Scene, for instance, here are Horatio's lines in their entirety:

> Custom hath made it in him a property of easiness. (5.1.67)
>
> It might, my lord. (81)
>
> Ay, my lord. (87)
>
> Not a jot more, my lord. (113)
>
> Ay, my lord, and of calf-skins too. (115)
>
> What's that, my lord? (196)
>
> E'en so. (199)
>
> 'Twere to consider too curiously, to consider so. (205)

As these characterless remarks suggest, Hamlet is engaged in a sustained soliloquy on death that is given the semblance of dialogue by the monosyllabic responses of his friend. The only authentic dialogue that takes place in this scene is between Hamlet and the Gravemaker, whose brief appearance here, like that of Osric in the following scene, brings before us a flesh and blood presence (however preposterous in Osric's case) that underscores by contrast the insubstantiality of Horatio.

TWENTY-FIVE

Self-Erasing Messengers: Go-Betweens and Get-Betweens

LEAVING HORATIO poised almost on the edge of extinction, let us expand a bit on the theme of dissolving middlemen. By means of Claudio and Horatio we have suggested that name, personality, individuality—whatever combines to make someone an end in himself—stands in inverse proportion to function, mediation, instrumentality—whatever reduces someone to a means to other ends.[6] The more Horatio becomes a perceptible function of Shakespeare's dramaturgy the less he seems a real person in Denmark. In fact, in acts of communication this self-erasure is the goal of the messenger, who seeks to convey his message so efficiently that he himself disappears from the sequence "Sender – Messenger – Receiver" and what remains is simply direct discourse between Sender and Receiver. Yet the go-between messenger, no matter how perfect at his work, is also a blocking get-between, a frictional presence that cannot wholly disappear. Moreover, even if the messenger could disappear, the message itself would remain as functional intermediary. Even face-to-face dialogue must make use of speech which, however self-effacing, constitutes a breathy verbal impediment to communication when compared, say, to the wordless "intelligence" by which angels simply understand, or to the silent communion of lovers' minds. Thus, just as the excess of messengers bringing Hamlet's letter to Claudius would seem to hinder its conveyance, the language of the message itself is a deterrent to understanding: "What should this mean?" the King asks, and, baffled, repeats aloud, "Naked! . . . Can you devise me?" To which Laertes, "I am lost in it, my lord" (4.7.48 ff.).

In *Hamlet* Shakespeare seems uncommonly conscious of the paradoxes of mediation, whether it takes the form of agents, actions, or words. In fact, he seems to have become preoccupied with what are now called "recursive processes," although the nearest he comes to expressing the notion is in the remark of Polonius about finding out directions by means of indirections (2.1.63). Recursive processes involve the interruption of a major action by one or more minor actions of a similar but usually simpler sort.[7] The Inner-Play, "The Murder of Gonzago," is a perfect example of a recursive structure. Similarly, the minor message-passings between the Sailor and Claudio and between Claudio and the Messenger are recursions within the major act of communication between Hamlet and Claudius. With this in mind, let us look again, for instance, at the curiously roundabout manner by which Fortinbras makes his final entrance into Denmark. When the direct action of his intended invasion of Denmark is intercepted by Old Norway, Fortinbras obediently turns his attention to a substitute invasion of Poland. Not all of Poland. Merely a "patch of ground" within Poland that, as the Norwegian Captain tells Hamlet, "hath in it no profit but the name" (4.4.18–19). And what is this name which constitutes the sole virtue of the place? We are never told it. The absence of this all-important name seems as puzzling here as the presence of a name for the wholly insignificant Claudio does in 4.7. Although the existence of a name is advertised, in effect the "patch of ground" remains as mysterious and unlocatable as Claudio. And yet the conquest of this piece of Poland, which seems at least irrelevant and at most an impediment to Fortinbras' hopes in Denmark, turns out to be a symbolic displacement of his original invasion of Denmark, which now seems not to have been abandoned so much as deferred and redirected. Thus from a blocking get-between the Polish invasion is unexpectedly transformed into a bridging go-between that enables Fortinbras to achieve all he had formerly sought in Denmark. In the process, the name of the patch of ground disappears, as if absorbed by the functional role of the place in Fortinbras' indirect search for a way to recover his father's lost lands.

Another instance of this "multiplication of the middle" by Shakespeare—his making a direct process most mediately indirect by means of recursion—lies in his engineering of the meeting between the Ghost and Hamlet in Act 1. This is a recursive process because the major meeting is postponed by various minor meetings—the Ghost with the guards, the Ghost with Horatio—and because the purpose of that meeting, the Ghost's conveyance of its message to Hamlet, is deferred by various minor message-passings—from the guards to Horatio, from Horatio to Hamlet. Here again the go-between/get-between paradox of mediation is apparent. For once Hamlet and the Ghost are within earshot, the intermediaries, in whose presence the Ghost refuses to speak, become not merely superfluous but a positive hindrance to communication. Their role is summed up in action when as the Ghost beckons Hamlet to follow it they intervene:

> *Mar.* You shall not go, my lord.
> *Ham.* Hold off your hands!
> *Hor.* Be ruled, you shall not go. (1.4.80–82)

The transilience from enabling go-between to detaining get-between is neatly compassed in the ambiguity of Hamlet's word "lets" when he wrenches free crying, "By heaven, I'll make a ghost of him that lets me!" (85) This threat to make a ghost of his friends is figuratively fulfilled after the departure of the Ghost when he tells them, "For your desire to know what is between us,/ O'ermaster it as you may" (1.5.140–141). "What is between" Hamlet and the Ghost is no longer his friends, whom he erases from the communicative process by refusing to confide in them, but the ghastly message itself. The function of that message is something we shall have to return to a little further on. First, let us examine an instance in which the middle is again multiplied and then erased, but also in which the message itself is transformed.

When Claudius ships Hamlet to England, Rosencrantz and Guildenstern act as messengers carrying the royal commission to England. The secret message of that commission is, as Claudius bluntly phrased it earlier:

> Do it, England,
> For like the hectic in my blood he rages,
> And thou must cure me. (4.4.69–71)

This is the message that Hamlet intercepts at sea, erases, rewrites, and sends on by means of the original unwitting messengers. As a result, a three-part process with Claudius as sender, Rosencrantz and Guildenstern as messengers, and England as receiver is expanded into a five-part process, thusly:

Claudius – R & G – Hamlet – R & G – England.

This transformed sequence parallels that of the "Claudio redundancy"—the obvious difference, however, being that whereas Claudio merely passed on his message, Hamlet does a very great deal in the way of erasure and revision before passing his on. Yet Hamlet must suffer the same fate as Claudio, inasmuch as the success of his stratagem depends upon this five-part process reverting to a three-part one again. That is, Hamlet himself must disappear from the sequence—even more completely than Claudio from his—leaving neither self nor name behind. His success in this role of self-erasing messenger is attested to by the subsequent deaths of Rosencrantz and Guildenstern, which are reported, appropriately, by an "Ambassador" at the end of the play.

TWENTY-SIX

Clearing the Middle

IN THESE several cases we have seen a recursive multiplication of the middle and an erasure of the "personality" of the middleman by the function he performs. To preserve itself, it seems, functionality must guard against the encroachments of personality on the one hand and its own tendencies to excess on the other. It becomes most itself when least apparent. The task performed with flair and style ceases to be purely functional, just as the tool wrought with ornamental designs ceases to be merely a tool. Witness the tool Osric, whose ornamental language and flourishing manner are so arresting in themselves that Hamlet and Horatio can scarcely grasp the import of his message from the King, which is perhaps why a second messenger, an efficiently anonymous "Lord," arrives shortly after Osric's departure to clarify matters. Osric is a sorry messenger indeed, and yet because of that he acquires a permanent place in the roll-call of memorable characters. Thus function, "growing to a plurisy,/ Dies in his own too-much" (4.7.117–118), and go-betweens have a fatal tendency to become get-betweens.

When Osric becomes too great an impediment, he is simply replaced by someone less distracting than he. The ultimate aim of messengers, of mediation as such, is, as we have observed, to become self-effacing, ideally to become self-erasing and thus to eliminate the middleman and bring about a direct confrontation between the principals. This reversion from mediacy to immediacy is a central movement of the play itself if we consider its major action as a contest between Hamlet and

Claudius. It has often been noted that the paradigm for their engagement is the duel, to which Hamlet himself calls attention in remarking on the deaths of Rosencrantz and Guildenstern:

> 'Tis dangerous when the baser nature comes
> Between the pass and fell incensed points
> Of mighty opposites. (5.2.60–62)

As Hamlet's remark indicates, the space between these two duelists is far from empty. "Between the pass and fell incensed points" of Hamlet and Claudius come, in order of their appearance, Polonius, Rosencrantz and Guildenstern, Ophelia, Gertrude, and Laertes, each in his turn "sent" by Claudius either to sift Hamlet's intentions or to kill him. As agents and tools of the King these characters surrender their own personality and integrity to their functional role, as is most evident in Guildenstern's abject willingness to do the King's bidding.[8]

> But we both obey,
> And here give up ourselves in the full bent
> To lay our service freely at your feet,
> To be commanded. (2.2.29–32)

Or as Hamlet says after his sea-voyage, referring to their imminent deaths, "Why, man, they did make love to this employment" (5.2.57).

To return the contest to its properly unmediated form, then, Hamlet must clear the overpopulated space between him and Claudius. His means are erasure and revision. Polonius, Rosencrantz and Guildenstern, and Laertes are erased; and Ophelia and Gertrude are, like the royal commission, "revised"—the one directed to a nunnery safe from pollution, the other exhorted to sexual abstinence.

If Hamlet finds his way to Claudius blocked by get-betweens, Claudius is similarly frustrated. In fact, the King's need for agents is occasioned by Hamlet's having interposed his "antic disposition" between his real self and Claudius. Claudius uses people, Hamlet invents them, as roles for himself to play. And part of that play is wordplay, in which the go-between of speech is so compacted by ironies, ambiguities, and figurations as to become an impenetrable barrier to communication. At this point Claudius sends agents not to interrogate Hamlet but to kill him.

In the last act, however, the thickened center of mediation is dissolved. Hamlet has set aside madplay and wordplay and stepped forward as himself. He has used words as daggers to dispatch Rosencrantz and Guildenstern. Now he uses the unbated dagger itself to eliminate Laertes, the last of Claudius' functionaries, and to encounter the King directly. The "interim" that Hamlet once claimed as his own (5.2.73) is now truly in his possession as a space is at last cleared for his fatal swordthrust.

TWENTY-SEVEN

Filling the Middle

THIS ERASURE of get-betweens accomplished by Hamlet does not, however, take account of the greatest get-between of them all, Hamlet himself, for that is his role in the major act of communication in the play. In it the Ghost is the sender, Hamlet the messenger, and Claudius the receiver. The message of course is "revenge." In this case, though, the message is implicit in the messenger. At the moment when Hamlet kills the King he will identify himself as the revenger, his action conveying the message, "Here, your Majesty—death! From my father." However, like most messengers of death in revenge tragedy, Hamlet is somewhat late with his delivery, and in fact is so conspicuously self-conscious about his lateness that he appears to doubt the validity of his message and his own suitability as its carrier. Hamlet the self-seeker who makes the readying for revenge an end in itself intercepts Hamlet the messenger of death and erases his message by erasing his identity. He does not rewrite the message immediately, as he did aboardship. The rewriting occurs rather at the moment of delivery when Hamlet conveys to Claudius a double message: "Death, from my father, and from me." The revised process takes the familiar five-part form:

$$\text{Ghost} - \text{Hamlet}_1 - (\text{Hamlet}) - \text{Hamlet}_2 - \text{Claudius}.$$

Here again the middle is multiplied, with an apparent foregrounding of medial function and, one would expect, a corresponding erasure of personality in the functionary. In this case, however, the bracketed "(Hamlet)" is at the center of the sequence not because he is a messenger but because he is a nonmessenger. Whereas Claudio disappears into his function because he performs it so invisibly well, because he is no more than that function, Hamlet acquires prominence because he performs his messenger of death function so badly, because he is so much more than that function. Claudio's personal presence in the play consists entirely in his possession of a name. Hamlet, on the other hand, takes on enormous personal presence in the play by forgoing his function and the name it entails (the name that would identify him as son, not self) in favor of a complex attempt to earn his own identity and name. This conversion of Hamlet from go-between to get-between by a foregrounding of his personality is mirrored in his own behavior, as his wordplay makes figuration an impediment to communication in the court, and his madplay interposes a barrier of "acting" between his real self and his Danish audience.

Thus the messenger "Hamlet_1," who is conveying death to Claudius, is intercepted by "(Hamlet)," who erases that message and function as he becomes the self-probing truant. Then this "(Hamlet)" is transformed into the final messenger "Hamlet_2," who conveys the rewritten message of death to Claudius in the Duel Scene. The question naturally arises, "How does '(Hamlet)' become 'Hamlet_2'?" Such a transformation would necessarily involve a sacrifice of personality to function, as we noted earlier when Hamlet intercepted and rewrote the King's commission during his sea-voyage. By that time Hamlet has in large part established his identity, found his name, and come to terms with his own imagined death. But this achievement of personal identity must be momentarily suppressed in the interests of self-preservation. Thus when he rewrites the commission he is obliged to adopt the identity of the King and to engage in an act of self-erasure by disappearing into his secret function. This loss of individuality is the price that must be paid sooner or later if Hamlet is to accept the role of revenger demanded by the Ghost and by the genre of revenge tragedy. Hamlet begins to pay that price during his sea-voyage and makes his final payment in the Duel Scene.

Such a payment carries its share of risk. I argued in Part I that the first of Hamlet's two killings of the King helps rescue the wandering middle of the play from structural irrelevance by making it consequent upon his earlier self-discoveries. A similar point emerges when we regard revenge as the completion of a communicative sequence that has been suspended by the truancy of "(Hamlet)." The danger in completing the sequence is that the individualized "(Hamlet)" will be depersonalized by his murderous function and thus disappear like all other medial messengers in the play. The uniquely multifarious self that has come so eloquently into the fullness of dramatic life by pondering on death (and love, honor, fatal flaws, betrayal, etc.) now seems about to commit suicide by extinguishing itself in a generic, functional role.

But of course Shakespeare is too fond of Hamlet to permit that to happen. Nor has Hamlet himself resisted such a self-obscuring role for so long merely to submit abjectly to it now. If he reaffirms his willingness to act as Death's messenger, his willingness takes the form, not of blind violence as when he killed Polonius, or of diabolic plots such as Claudius is himself devising, but of "readiness" within a providential context that will itself supply occasions to act. Thus when the moment of delivery arrives, the situation itself not merely rewrites the message to read "I am my own and my father's avenger" but in doing so defines Hamlet as a joint sender of the message as well as its messenger. Hamlet is not merely his father's agent, a functional nonentity like Claudio, but the conveyor of his own meanings, an actor in his own behalf. In fact, given the deadliness of the poison he has already administered to the King with the unbated sword, Hamlet's second murder of Claudius is functionally superfluous. It is a mere formality, a gesture of deference to his father by a son who has proven himself his own man.

But if Hamlet is to be fully rescued from self-erasure his story must be told. "Things standing thus unknown," he must seem no more than an agent of destruction to those "that are but mutes or audience to this act" (5.2.348, 337). Thus one more message must be sent, from Hamlet to the "yet unknowing world" by means of Horatio, the speaker. But in that case Horatio himself, whom we last saw as a mere adjunct of Hamlet and agent of Shakespeare, will be in need of rescue. As for his story, it has an even greater function than that of restoring Hamlet's "wounded name" (5.2.347)—but that belongs at a further stage of the discussion.

TWENTY-EIGHT

Diachronic Suicide

WE LEFT Horatio suspended on the verge of a kind of suicide. The more he becomes a function of Shakespearean theater, Hamlet's messenger to the Globe, the more he approaches extinction as a real person in Elsinore. Other messengers and message-revisers—Claudio is the paradigm—are in like peril. Becoming nameless by being named, disappearing in the process of making their presence felt, they comply with the principle of "To be and not to be" that operates ubiquitously in *Hamlet*. This principle derives from verbal negation, the instrument by which language seeks and fails to commit suicide. Like the negative, the various messengers in *Hamlet* are in the awkward situation of Oberon in *A Midsummer Night's Dream*, who in full view of the audience is obliged to claim, "I am invisible" (2.1.186). This kind of suicidal unvanishing act comes more easily, one assumes, to fairy kings than to Danish princes; and yet the impulse to self-slaughter is everywhere in *Hamlet*—in the Prince, in Horatio, in Ophelia, in the language of the Prince, in the functionality of the play's messenger service. More than that. If various processes of communication within the play are prone to self-erasure, what about the master communication, the play itself? Is there a theatricidal *Hamlet* that corresponds to the suicidal Hamlet? In a sense all plays are theatricidal, but few playwrights know this well enough to incorporate their knowledge into the form of their work. To get a notion of how Shakespeare has done this in *Hamlet*, let us glance first at some features of language made familiar by various linguistic theorists.[9]

When Jonathan Swift said "Proper words in proper places, make the true definition of style," he had reference to two modes of arrangement that are operative at all levels of language from phonemes to discourse. They are usually called the vertical dimension, with which are associated *langue*, synchrony, paradigm, and metaphor, and the horizontal dimension, with which are associated *parole*, diachrony, syntagm, and metonymy. The vertical dimension features acts of selection, the horizontal features acts of combination. Roughly put, for example: in fashioning the sentence quoted above, Swift would select his "proper words" from a field of verbal possibilities—a synchronic, paradigmatic, vertical vocabulary—and then combine them in their proper diachronic, horizontal places in accordance with the rules of grammar and syntax. That is, Swift would choose his opening word "Proper" from other adjectival candidates sufficiently similar that they might serve as substitutes: terms like "correct," "appropriate," "suitable," "right," "fitting," and so on. Having made this particular (paradigmatic) choice of an adjective, he is now obliged (syntagmatically) to combine it at some further point with a noun, to be selected from such vertical possibilities as "words," "terms," "expressions," "diction," and "locutions." Then that noun, functioning as a subject, would imply the imminent selection of a verb, and so on.

From this standpoint a work of literature could be regarded as a sentence whose words (agons, scenes, episodes, chapters, stanzas) are chosen from a vertical vocabulary of possibilities and combined horizontally according to the relevant rules of literary syntax (story, plot, structure, generic form).[10] A further development appears in Roman Jakobson's claim that metaphor and metonymy (including synecdoche) belong to different linguistic axes, although they have been traditionally regarded as close figurative relations. Metaphor is suited to the vertical axis, he says, because in it the vehicle is selected to substitute for the tenor by virtue of their similarity, however remote. But when we substitute metonyms like "White House" for the President or "hands" for a crew of workers, we do so, not because they are similar, but because they are contiguous (somewhere in the nearby spatial or conceptual vicinity). Jakobson then claims that a discourse may be "forwarded" on either a metaphoric or a metonymic basis, depending on whether one topic leads to another through their similarity or through their contiguity, and he suggests that within literary discourse "for poetry, metaphor, and for prose, metonymy is the line of least resistance."[11]

In using terms like "line of least resistance" and the "forwarding" of discourse, Jakobson stresses the diachronic mobility of language, and with good reason. If language is to mean, it must move, toppling its static vertical vocabulary over into kinetic horizontal combinations. For the poet, however, this fall into syntax may seem at times a heavy price to pay for meaning, and, like Falstaff owing God a death, he may be loathe to pay before his day. Why this should be so, and why it is especially so in *Hamlet*, will become apparent if we consider Hamlet's attitudes toward the diachronic, which he has good cause to distrust.

Had the normal syntax of royal succession prevailed, the Danish crown would have passed metaphorically from father to son—"metaphorically" because of their likeness, as implied by their shared name. But this process has been aborted and diverted by Claudius. The transfer of the crown from Old Hamlet to Claudius, though from brother to brother, is not metaphoric but metonymic: "My father's brother, but no more like my father/ Than I to Hercules" (1.2.152–153). Diachronic affairs would have taken place, as in a world of time they must, even without the intervention of Claudius. But his murder of old Hamlet, his swift accession to the throne, and his o'erhasty marriage to Gertrude both accelerate and pervert the normal orderly process of events in Denmark. Hence Hamlet's lamentations about the unceremonious passage from funeral to marriage in his first soliloquy, and the caustic ironies of his "Thrift, thrift, Horatio! The funeral baked meats/ Did coldly furnish forth the marriage tables" (1.2.180–181). Ceremonies—normally the hinges of social experience that articulate the meaningful—have fused in Elsinore to produce the grotesque oxymorons of Claudius' opening speech about "mirth in funeral and dirge in marriage" (1.2.12).[12]

That Claudius should be responsible for the jamming together of opposites and dissimilars is appropriate to his metonymic role. For if the metaphoric pole features the selection and substitution of similars, what would not appear along its axis are opposites and unlikes, which fall along the metonymic axis. Hence the oxymorons of Claudius' opening speech, and hence too Hamlet's (metaphoric) likening of the succession of Claudius to the (metonymic) transformation of "Hyperion to a satyr" (1.2.140). Moreover, as befits the kinetic aspect of metonymy, Claudius shows himself in the Council Scene to be a king who goes about his royal business with great dispatch, passing item by item through the calendar of court affairs. And the business he deals with is motion. He forestalls the invasion of Fortinbras by sending ambassadors to Old Norway, he permits the departure of Laertes for France, and he detains Hamlet from returning to Wittenberg. But if he restrains Hamlet's geographical movement, Claudius is most anxious that the Prince move emotionally from grief toward resignation through an acceptance of the inevitable diachronic facts of life, and death:

> But you must know your father lost a father,
> That father lost, lost his, and the survivor bound
> In filial obligation for some term
> To do obsequious sorrow. But to persever
> In obstinate condolement is a course
> Of impious stubbornness. (1.2.89–94)

Claudius understands as well as Hamlet does the deceitfulness of time—

> But that I know love is begun by time,
> And that I see, in passages of proof,
> Time qualifies the spark and fire of it (4.7.111–113)

—and he knows as well as Macbeth does how fugitive are time's invitations to the will:

> That we would do,
> We should do when we would; for this "would" changes
> And hath abatements and delays as many
> As there are tongues, are hands, are accidents,
> And then this "should" is like a spendthrift's sigh,
> That hurts by easing. (4.7.118–123)

Claudius himself is not one to suffer abatements and delays. He sees risks, accepts them, and acts. He is not hesitant to begin nor, once begun, to continue. He pauses once, in the Prayer Scene. Even then, however, anxious to keep what he has achieved—"My crown, mine own ambition, and my queen" (3.3.55)—he does not long for a return to lost innocence but for a going-on without penalty: "May one be pardoned and retain the offence?" (3.3.56) For Claudius, as for Macbeth, returning is as tedious as go o'er.

Returning, in fact, is something of a problem for Claudius—not merely his own return to innocence, which he cannot manage, but also the returns of others. As the King he controls departures. To become king he has "dispatched" his brother, and during the play we see him dispatch ambassadors to Norway, Laertes to France, Rosencrantz and Guildenstern to spy on Hamlet, Polonius to do the same, Fortinbras through Denmark to Poland, and Hamlet to his intended death in England. What the King cannot control are return journeys. The Ghost returns unbidden to tell fearful secrets, Laertes to storm the palace, Hamlet to kill a king, and Fortinbras to become a king. Even his own poison returns to Claudius.

In the diachronic way of things, death is the ultimate departure, the last exit, which makes its first entrance in *Hamlet* in the form of Claudius' murder of the old king. Claudius ushers death into the play but, once it is there, Hamlet entertains it almost obsessively. Hamlet is repelled, it seems, not so much by death itself as by the diachronic degradations that lead both to and beyond the grave. If death is conceived as a synchronic suspension, as sleep ("to die, to sleep"), or as a "quietus" to a metonymic series of life's calamities ("The oppressor's wrong," etc.), then it is a "consummation/ Devoutly to be wished" (3.1.61 ff.) But if seen as a way station en route to undiscovered countries in which life's bad dreams get worse, then death is sufficiently appalling to turn awry the impulse to suicide and to high enterprise alike. Not merely death, then, but life's fallings off on the way repel Hamlet—all the low transformations:

> Why, she would hang on him
> As if increase of appetite had grown
> By what it fed on, and yet, within a month . . . (1.2.143–145)

> . . . the satirical rogue says here that old men have grey beards, that their faces are wrinkled, their eyes purging thick amber and plum-tree gum, and that they have a plentiful lack of wit, together with most weak hams. All which, sir, though I most powerfully and potently believe, yet I hold it not honesty to have it thus set down, for yourself, sir, shall grow old as I am, if like a crab you could go backward. (2.2.197–204)

> Ay, truly; for the power of beauty will sooner transform honesty from what it is to a bawd than the force of honesty can translate beauty into his likeness. This was sometime a paradox, but now the time gives it proof. (3.1.112–116)

And the even lower transformations that take place once one gets to the grave:

> Your worm is your only emperor for diet. We fat all creatures else to fat us, and we fat ourselves for maggots. Your fat king and your lean beggar is but variable service, two dishes, but to one table—that's the end. (4.3.21–25)

> Alexander died, Alexander was buried, Alexander returneth to dust; the dust is earth; of earth we make loam; and why of that loam, whereto he was converted, might they not stop a beer-barrel? (5.1.209–212)

These metonymic progresses through the wormy earth, which will take a king "through the guts of a beggar" (4.3.32), are not so lofty as Gertrude would have it when she says "all that lives must die,/ Passing through nature to eternity" (1.2.72–73).

Hamlet's gross biological reversals from life to death, from flesh to dust, and from feeder to fed-upon ("Not where he eats, but where 'a is eaten," [4.3.19]), provide an ironic frame for all processes in *Hamlet*, especially human actions. Thus the play abounds with instances of what Horatio calls "purposes mistook/ Fallen on the inventors heads" (5.2.386–387), instances in which, as Thomas F. Van Laan puts it, "an individual's self-assertion brings him only self-destruction."[13] These range from the death of Old Fortinbras in combat through the deaths of Polonius, Rosencrantz and Guildenstern, and Laertes to that of Claudius himself, each of whom is in some sense an "engineer/ Hoist with his own petar" (3.4.213–214). As Van Laan observes:

> Since the victims of ironic reversal in this play range from the guilty through the neutral to the innocent, the conclusion must be that the universe of *Hamlet* negates all human activity, whether vicious or not.[14]

If the universe of *Hamlet* is inimical to human activity, afflicting even the most straightforward of acts with peripeties, degeneration, and death, then how does Hamlet stand, whose role requires of him the act of revenge? The perplexity of his plight is suggested if we regard revenge as an attempt to erase diachronic change and thereby symbolically to restore the past. Through its likeness to the original murder (catching Claudius, like Old Hamlet, "full of bread"), Hamlet's revenge would superimpose itself on and by a kind of sympathetic magic eclipse the crime against his father. But this attempt to nullify the diachronic, like Tristram Shandy's effort to record his own history, can only be made by an act that inserts

Hamlet into the diachronic process itself, polluted as it is in Denmark by its peculiar Claudian contaminants.

Faced with this enigma, Hamlet finds a temporary solution not in action but in a withdrawal from action into "acting." His "antic disposition," composed of "actions that a man might play," offers him the illusion of the diachronic within the bracketed safety of the synchronic, a way of acting without acting. But madplay is an insecure refuge and an ineffectual mode; such "acting" casts no shadows and certainly kills no kings. Summed up in syntactic terms, Hamlet at this point is like the subject of an as yet unpredicated sentence. He has drawn about him a wealth of self-modifying adjectives, phrases, and clauses that enrich his being and subtilize his character. But all of this substantive ornament is massed and suspended in time until Hamlet commits himself to the transitive verb prescribed by the Ghost. That act of predication, however, will thrust Hamlet into his world and time, into the diachronic current that goes its way indifferent to the irreplaceable "I" and empties at last into a common grave.

TWENTY–NINE

THEATRICIDE: DRAMA AS FUNCTION

EARLIER we left Horatio at the point of suicide, about to vanish into his dramaturgical function as Hamlet's messenger to the Globe, and now we leave Hamlet in a like state, poised for an act that will consume both his distinctive personality and his life. The play, it seems, is fraught with suicidal intendments. Hamlet laments that the Everlasting has "fixed/ His canon 'gainst self-slaughter" (1.2.131–132). Horatio proclaims himself "more an antique Roman than a Dane" and seizes the poisoned cup (5.2.343). Ophelia goes to a watery death by mad accident or mad design. Laertes leaps into his sister's grave and cries to the gravediggers, "Now pile your dust upon the quick and the dead" (5.1.251). And even the much-protesting Player Queen calls for suicidal punishment should she prove false: "Nor earth to me give food, nor heaven light" (3.2.214). Then there are the metaphoric suicides of the various messengers who in accepting their tasks erase their selves. And finally there are all of the actions characterized by "ironic reversal" whereby self-assertion proves a form of self-destruction.

Given this abundance of suicidal modes, we may wonder if Shakespeare were not also conscious of the suicidal impulse within drama itself. Perhaps his uncharacteristically topical mention of the War of the Theaters suggests as much. Of course the internecine struggle between the common players and the child actors is itself a form of theatrical suicide. But, in addition to that, consider the "little eyases" themselves, of whom Hamlet asks,

> Will they not say afterwards, if they should grow themselves
> to common players—as it is most like, if their means are
> no better—their writers do them wrong, to make them exclaim
> against their own succession? (2.2.346–351)

Within the irony of the child actors attacking their own future as common players, like freshmen deriding sophomores as such, is the diachronic curiosity of their committing professional suicide by growing up. They will serve physiologically as their own executioners in the process of aging—when, for instance, their voices change: "Will they pursue the quality no longer than they can sing?" (346–347). Like all children, the "little eyases" must use themselves up, erasing what they are at each point in order to become what they are to be.

That is what Hamlet finds most repellent about the diachronic. Each phase of the biological process, for instance, becomes merely a means to another end, which in its turn becomes a further means to further ends:

> Your worm is your only emperor for diet. We fat all
> creatures else to fat us, and we fat ourselves for
> maggots. (4.3.21–23)

As each phase of the process fulfills its transient function it is used up and forgotten. Thus Gertrude's passage from one husband to another dismisses Hamlet's father as merely a phase of her overall marital career, a phase all too readily forgotten: "O God, a beast that wants discourse of reason,/ Would have mourned longer" (1.2.150–151). Or as Claudius says, perhaps having himself begun to sense a falling-off in Gertrude's love for him ("I see, in passages of proof" [4.7.112]):

> There lives within the very flame of love
> A kind of wick or snuff that will abate it. (4.7.114–115)

If life and love are suicidal, what of such flickering things as plays? Insofar as each phase of a play is just that, a phase, its function is to advance the action to the next phase of the plot; and its fate—like the fatted body in the grave, like Gertrude's first husband, like love—is to be abandoned in the wake of its own progress. The more efficiently a given scene conducts to its successor the more its own distinctive nature succumbs to its functional role. To revert to the metaphor of communication, the sequential phases of a play are the self-erasing messengers of the play's own message. To the extent that all of the parts perform their diachronic roles, the play becomes self-consuming in becoming self-fulfilling. It achieves its full dramatic life only through a series of theatricidal acts. This paradox is in accord with Hamlet's remark about the sleep of death terminating the heartaches and shocks of life: "'Tis a consummation devoutly to be wished" (3.1.64–65). For this "consummation" is both an annihilation and a completion (a consummating): a self-consuming fulfillment and a fulfilling self-consumption.[15]

One could make such a claim of any literary work or lineal structure of meaning, but drama in performance is peculiarly vulnerable to self-consumption. Whereas written works preserve their own past moments physically on a page, the words and actions of the theater dissolve at the instant of their presentation.[16] Shakespeare can speak confidently enough of the durability of his printed verse:

> So long as men can breathe or eyes can see,
> So long lives this, and this gives life to thee. (*Sonnet 18*)

> Not marble nor the gilded monuments
> Of princes shall outlive this powerful rhyme. (*Sonnet 55*)

But the life of theatrical speech rides on breath itself, exhalations of eloquence that cannot survive their own saying; and the grand spectacles of the stage are transient as a gesture—as Prospero's gesture dismissing the revels. This transiency issues in part from the present-tense mode of drama—a tense whose presence is constantly disappearing into a past that exists only as a fading memory in the minds of an audience now intent upon a new presence.

Actors, as Hamlet the director is keenly aware, resent this self-consuming bent of theater. They will try, by mouthing their lines, he says, or sawing the air too much, to draw attention to themselves, to become show-stoppers—especially the clowns, who will angle for laughs "though in the mean time some necessary question of the play be then to be considered." "That's villainous," Hamlet declares, "and shows a most pitiful ambition in the fool that uses it" (3.2.42–45). This, of course, from a Hamlet who, while playing the fool and jester in court, has deferred the act of revenge which even he regards as "a necessary question of the play."

One would expect not merely the players but the playwright himself to strive against this gradual dying of the play as it advances toward its fulfilling demise. For if each phase of *Hamlet* were to become merely a disappearing messenger in the overall communicative act of drama, a way station of meaning toward which the action moved with anticipation and from which it departed with spent interest, then the play would be but a melodramatic thriller or the simplest sort of didactic allegory. The theatrical and critical history of *Hamlet,* however, offers little support to the view that the play conveys a single message unaltered from beginning to end and that its parts disappear into their functions. Like Hamlet with the King's commission at sea, each part of the play transcends its own functional death by rewriting the message it receives. Each part demands its say in the dialogue of meaning, its assertion of independent worth apart from transient functionality. Each phase—word, sentence, scene, act, play itself—rages, albeit quietly, against the dying of its own light. And yet it goes without saying that the light will die, that time will have its way with *Hamlet* the play as it does in the end with Hamlet the prince. But in the meantime?

THIRTY

Naming of Delay/ Naming as Delay

WE HAVE got to the inauspicious point where Horatio, Hamlet, and now the play itself are verging on suicide and a descent into the grave of functionality. The question is, What can Shakespeare do to rescue them? And the answer from one point of view, is Nothing. Hamlet is doomed to die within the play, and the play to expire by self-"consummation." As for Horatio, he is rescued by Hamlet from death within the play, but only to die with it, as part of its general consummation. If it be not now, yet it will come. But if the playwright cannot prevent the arrival of death, perhaps he can at least delay it.

Delay, after all, comes with the generic territory. It is simply "given" that in revenge tragedy the initial crime should be separated from the terminal revenge by a period in which the action advances without progressing. This apparent defect of structure Shakespeare converts into a virtue by invoking a kind of large-scale version of what Jacques Derrida calls, in reference to the differing/deferring character of signs, "*differance.*"[17] That is, as I argued in Part I, both Hamlet and *Hamlet* defer the act of revenge by interposing between beginning and end a dramatic middle in which the hero distinguishes himself from others, from his own past self, and from his potential selves (by means of the negative mode described in Part II), and in which the play distinguishes itself from its past identity as the *Ur-Hamlet* and from other revenge tragedies (by means of a self-referential exploration of its own form). Or, as we have been phrasing it in the present chapter, the delivery of the Ghost's message of revenge, which requires Hamlet to submerge himself in the function of messenger, is impeded by the mid-play stress on the personality of both hero and play in their joint role as nonmessenger.

Thus this period of unavoidable structural delay is devoted to the naming or individualizing of the hero and the play. More than that, however, it is also devoted to the naming of delay itself. Instead of attempting to obscure the hiatus in his dramatic structure, as many revenge plays do, Shakespeare makes a point of publicizing its presence. As William Empson puts it, when Shakespeare took a careful look at the structure he had inherited he must have thought,

> "The only way to shut this hole is to make it big. I shall make Hamlet walk up to the audience and tell them, again and again, 'I don't know why I'm delaying any more than you do; the motivation of this play is just as blank to me as it is to you; but I can't help it.' "[18]

This naming of delay by Shakespeare is like his naming of Claudio: unnamed, both would have disappeared into their functions, and we should never have known of their existence.

But why should Shakespeare want to advertise the presence of a structural defect inherent in revenge drama? In part, I suggested earlier, in order to transcend that genre by incorporating the very issue of generic form into the content of his play, thereby making that form not merely a means of representation but an end in itself, an object of representation, and thus giving his own play a differentiating individuality. But also, I think, in order to subject to exploration the curious phenomenon of delay, which is a kind of positive negative, like Hamlet's antic disposition or Claudius' kingship. Hamlet plays mad, Claudius plays king; each role asserts "I am not what I am." But these histrionic negations are less elusively self-denying than Hamlet's delay, because we can define what is required for the role of madman (strange dress, wild and whirling words, distracted manner) and for the role of Player King (royal costume, stately speech, judicious air), but there is no prescription for the role of delayer. He may dress, speak, and act in any way he likes as long as it does not conduct him to the one thing he is delaying doing. As a result, whatever the delayer does do is nullified by reference to what he is really doing—to wit, nothing. Everything Hamlet says and does—his baiting of Polonius, his welcoming of the players, his soliloquizing, his discoursing on styles of acting, his reviling of Ophelia and his mother—is from the standpoint of his murderous mission irrelevant and self-erasing. The moment the members of the audience recall that mission—and if they do not recall it, the play itself will nudge them to do so—whatever Hamlet is doing at that time will dissolve into a version of not-revenge. At the same time, not-revenge materializes in the form of whatever Hamlet is currently doing.

In thus naming delay, then, Shakespeare imparts presence to absence and absence to presence, not, I think, as a solemn ontological affair but in a spirit of playfulness. After all, delay is itself a form of play—a diversion from the functional business of revenge—and as such it invites us to enjoy it for its own independent sake, even as Hamlet himself takes a certain perverse puritannical satisfaction in wandering off the highway toward murder in order to explore the swamps and sleughs of evil in Denmark. And although Hamlet himself may sometimes seem in danger of not finding his way back to that highway, we need not fear that Shakespeare is lost. His very advertisement of delay assures us that he is fully conscious of his generic commitment to revenge and, as an implicit promise to fulfill that commitment, offers us the freedom to divert ourselves in the meantime in the caesural something-nothingness between crime and revenge.[19] The act of revenge is suspended, and the self-consuming progress of the play is suspended, so that Hamlet can range in his imagination through the vocabulary of the possible before committing himself to the one preclusive act he is obliged to perform. Our functional interest in "what is going to happen" is supplanted by our intrinsic interest in the eloquent and multifaceted hero. The absence of revenge does not signal a failure of the hero so much as the success of the playwright who has transformed a negative nothing into a positive Nothing, giving not-revenge a compelling fascination. The absence of revenge entails the rewarding presence of self, of Hamlet at play, going nowhere except into his own mind and language, into a brown study of his world.

Shakespeare's naming of not-revenge enables us to witness once again the power of the negative to affirm the negated, to make a something of nothing. Let me take a smaller instance of absence acquiring presence by virtue of being named—Hamlet's lines about pausing in the "To be or not to be" speech:

> For in that sleep of death what dreams may come,
> When we have shuffled off this mortal coil,
> Must give us pause. (3.1.67–69)

In the theater this is usually delivered as though there were a question mark after "coil," so that "Must give us pause" becomes an independent sentence rather than the predicate of "what dreams may come." That may be syntactically wrong, but it is psychologically right, for the actual pause that readers and actors feel obliged to make at the conclusion of the sentence is induced by the expression "Must give us pause." It is as though "Must give us pause" were a stage direction incorporated into the monologue itself. Even within the phrase itself, the presence of the word "pause" arrests the normal rhythm of utterance, creating a juncture after "us" as though the reader had come upon an internal stop sign. Thus the naming of the idea of pause calls forth the phenomenon itself, the effect of which is to transform Hamlet's indecision about suicide from a simple absence of decision to a visible presence of not-deciding. The movement of the speech at this point, like the movement of the play during its period of not-revenge, can be characterized as dynamic stasis.

One more example: the First Player's account of how the descending sword of Pyrrhus stopped in the air as it "was declining on the milky head/ Of reverend Priam," so that

> as a painted tyrant Pyrrhus stood,
> And, like a neutral to his will and matter,
> Did nothing. (2.2.478–482)

The momentarily paralyzed Pyrrhus foreshadows Hamlet arrested in indecision over the praying Claudius and, more largely, Hamlet frozen in doubt and irresolution throughout the middle of the play.[20] But the moment of Pyrrhus' pause is brief. His sword resumes its bloody descent in time to catch the last life flickering in ancient Priam—and in his own good time, and Shakespeare's, even Hamlet at last makes an end of his vengeful business.

But let us reexamine Pyrrhus in pause. If we had only the action itself—if the killing of Priam were dramatized instead of narrated—the murderous sword would simply cease its descent and become for an instant a kinetic zero. In the Player's narration, however, the sword does not simply cease but rather "seemed i' the air to stick" (2.2.479). "Stick" is an example of what Murray Krieger calls *ekphrasis*, a species of unmoving motion: the sword is arrested in the air into which at the same time it thrusts.[21] This ekphrastic phenomenon is illustrated also by Pyrrhus himself, for despite the negative stress on his neutrality of will and physical inaction, his very motionlessness takes on caesural presence:

> But, as we often see, against some storm,
> A silence in the heavens, the rack stand still,
> The bold winds speechless, and the orb below
> As hush as death, anon the dreadful thunder
> Doth rend the region, so, after Pyrrhus' pause,
> Aroused vengeance sets him new a-work. (483–488)

Here is an epic simile whose images themselves crowd into the space of Pyrrhus' pause to create a verbal imminence that beggars the action to come and makes of motional nothingness a visible stasis.

From these examples it would appear that language, like nature, abhors a vacuum, since the very act of naming or expressing juncture constitutes a closure of junctural space. Thus a kinetic nothingness, an unmotion, acquires body and substance, as though the better to resist the momentum of the action it interrupts. To resist, however, is but a "standing still," not an overcoming. That is to say, juncture—or pause, hiatus, caesura—is by definition a medial entity, not a terminal one. The action it interrupts must resume again, thereby enclosing or bracketing the juncture, if the latter is to exist at all. Drama, it appears, will assert its essential nature as a performance in time despite the playwright's efforts to resist its destructive flow.

THIRTY-ONE

Forms of Arrest: Scene, Inset, Soliloquy

THE PAUSES we have just examined break the rhythm of time, and since time is death's agent, each of these pauses arrests a movement toward death: the death of Claudius by revenge, the death of Hamlet by suicide, the death of Priam by assassination. Long or short, the pause is inevitably temporary; death will make an end of Trojan kings and Danish princes and English plays, today, tomorrow, or on Monday. And death's end is traditionally definitive, the final summing up: "your fat king and your lean beggar is but variable service, two dishes, but to one table—that's the end" (4.3.25–27). And as with men, so with meanings—the meanings of sentences, for instance. Normally the linear advance of a sentence brings about a shrinkage of the range of its potential meaning, the shrivelings of age as it were. That is, a sentence moves toward a meaning that can be apprehended with certainty only at its completion, after we have filtered out of it the unintended but momentarily conceivable meanings that floated into consciousness while we were listening or reading—meanings cast up by the inherent polysemousness of language. In *Hamlet*, for instance, once we have become alerted to the prevalence of its puns we spend nearly as much time screening not-puns out of sentences as we do admitting punning meanings into them. Similarly, as we witness the play for the first time in the theater we are constantly entertaining and then dismissing various large-scale possibilities of its plot.

We may think, for example, that the play will, and then gradually realize that it will not, feature an invasion of Denmark by Fortinbras, or a courtship and perhaps marriage between Hamlet and Ophelia, or a swift revenge by Hamlet, or an Orestes- or Nero-like killing of Gertrude, or a full-scale rebellion led by Laertes, and so forth. As the plot proceeds, however, it erases its own potentials, those alternative roads it seemed for an instant to contemplate taking, and at the end it stands before us, not as what it might have been, but what it has proved to be.

From this standpoint, linear discourse, whether in the form of sentences or plays, is "tragic" in that it observes the Aristotelian recipe for the tragic plot: a causal sequence that reduces the possible to the probable, and the probable to the inevitable. Hence meaning—the tragic hero of this metaphor—must fall from what it grandly and variously seemed to be to what it sadly is, shedding like Lear its rich garments for nakedness. However, as Bernard Beckerman has argued, the plot of *Hamlet* is neither Aristotelian nor causative:

> What strikes many critics as a lack of unity in *Hamlet* is its particular pattern. Once the conditions imposed on Hamlet by the Ghost are revealed, we witness the following sequence: the place of love in Hamlet's mind, the testing of Claudius at the play, the relation of mother and son. Each event is prepared for, but each in turn gains full emphasis. Nor does one event bear causative relationship to another.[22]

The Nunnery Scene, that is, does not cause the Play Scene, nor the Play Scene cause the Closet Scene. The sequence is not causative but episodic. The Closet Scene itself, Beckerman points out, is most notably so:

> Certainly the scene is dramatic, in fact, one of the most dramatic in all literature. Yet it does not carry the action to a new stage, but allows Hamlet to express his disapproval and suspicion of his mother. In fact, the central portion of the scene leaves no trace on the plot.[23]

Or, in this connection, consider the Prayer Scene, which seems almost to announce Shakespeare's indifference to causation. The Play Scene has just supplied observable proof of the Ghost's veracity and Claudius' guilt, and when we last saw Hamlet he was full of stage revenger rant, talking about drinking hot blood and doing "such bitter business as the day/ Would quake to look on" (3.2.389–391). Enter, then, Hamlet primed for revenge into the presence of his vulnerable victim at prayer. But instead of presenting a logical culmination to this causal sequence, a scene of at least attempted revenge (Hamlet being interrupted, say, by the entrance of another character), Shakespeare fashions an inward moment in which the hero ponders the nature of the occasion and repudiates his opportunity as "mere hire and salary," as an act, that is, that would cast him in the demeaning role of functionary. By the same token, the scene itself, had it played its expected role in the pattern of cause and effect, should have become "mere hire and salary" also. But as it is, Shakespeare has arrested the functional process by means of Hamlet's explosive "No!" and substituted a scene devoted to the exploration of the consciousness of both hero and villain. The scene thus becomes semiautonomous, more an end in itself than either the effect of what has come before it or the cause of what will follow it. Resisting a purely instrumental role, it retains its own integrity and intrinsic interest.

Thus the "particular pattern" of *Hamlet* that Beckerman speaks of is not derived from the causal functioning of its plot. Its scenes are not erasable messengers in the long communicative act of drama but semiautonomous entities that arrest and defer the flow of action. Or as Mark Rose says:

> Each scene is, in a sense, a single speaking picture: proceeding through a play is somewhat like proceeding past a cycle of narrative paintings, pausing to absorb each picture in turn before stepping back to discover the unifying principles of the cycle as a whole.[24]

He is referring to Shakespearean drama in general, but this spatializing of the temporal takes on special significance in *Hamlet* because its varieties of artistic arrest—whether episodic in Beckerman's view, scenic in Rose's sense, or in other ways technical and stylistic—are metadramatic correlatives to the theme of delayed action. They correspond at the level of dramatic technique in the Globe theater to a host of interruptions, delays, intrusions, junctures, and indirections that impede the orderly course of time and human affairs in the Denmark of the play.[25] Fortinbras' invasion of Denmark is intercepted and diverted to Poland. Hamlet's schooling in Wittenberg is abruptly curtailed by his father's death. The love of Hamlet and Ophelia is frustrated by the interference of Polonius, whose later interruptions in Gertrude's closet are permanently cut short by Hamlet's sword. The presentation of "The Murder of Gonzago" is broken off by Claudius' cry for light. Hamlet's voyage to England is aborted by the encounter with the pirate ship. The fatal encounter of the "mighty opposites" is impeded by the intrusive presence of agents and intermediaries, lies, deceptions, and baffling wordplay. The beginning of the play features the intrusion of the Ghost come back from death, and the end features the intrusion of Fortinbras come back from Poland.

The genesis of this interruptive mode is the murder of Old Hamlet, which interrupts the king's journey not merely through life but also toward heaven. Caught in the middest, he finds himself marking time in a spiritual "prison-house" at the crossroads between heaven and hell. This ghostly pause of Hamlet's father broods like the Ghost itself over Denmark and Shakespeare's play. Until Hamlet can complete the act of revenge that will enable the Ghost, if not to complete its journey, at least to "rest" in some kind of peace (1.5.183), nothing in Denmark can properly conclude. Beginnings may be made, but their ends are inevitably deferred. Hamlet's proclamation "I say, we will have no more marriage" (3.1.149) confirms not merely the discontinuance of his courtship of Ophelia but the rift that divides all beginnings and ends in *Hamlet*. That rift can be closed, can be made an end of, it seems, only when Hamlet has himself seen the ultimate end in the graveyard and, accepting it for what it is, adopted a "readiness" for it. For played off against all these transient interruptions and incompletions in Denmark is the play's increasingly insistent stress on degenerative change conducting to death. That long slippage of life toward the grave is the one action whose completion is guaranteed.

But in the meantime life shows its deference toward death by deferring it, and Shakespeare's play resists its own dying—in little, by stylistic pauses of the sort we have examined earlier, in large, by semiautonomous episodes and tableaux-like scenes, and, in between, by what Francis Berry calls the Shakespearean "Inset." An Inset is an incident, usually narrated, in which "the imagined spectacle is at odds with the actual spectacle," as when the imagined spectacle of Hamlet, hatless and unbraced, finding his way without eyes from Ophelia's chamber, is at narrative odds with the actual on-stage spectacle of Ophelia conveying all of this to the wild-surmising Polonius (2.1.74–97).[26] The dramatic mode is conducted in the present tense and the indicative mood: in the here-now, this is happening. In the here-now of theatrical immediacy the Ghost addresses Hamlet, Hamlet greets his school-fellows, Polonius cries out from behind the arras, Hamlet lunges at the King, and Fortinbras bids the soldiers shoot. Most Insets, however, are narrative, and couched therefore in the past tense indicative: in the there-then, this happened. In the there-then, we are informed by various tellers, Old Hamlet killed Old Fortinbras in single combat, Claudius killed Old Hamlet in secret treachery, Young Fortinbras' invasion was diverted to Poland, mad Hamlet came sighing in disarray to Ophelia's chamber, mad Ophelia came in greater disarray to the willowed brook, and the seaborne Hamlet awoke, rewrote the King's commission, and boarded a pirate ship. The dramatic mode, by virtue of the continuity of its present tense, is one-dimensional, a single temporal plane. But the narrative mode is multidimensional, having at its disposal depths and striations of past time. It may depict events as shallow or near in time as the appearance of the Ghost or as deep and distant as the chivalric encounter of the old kings, or even the emptying of the graves in Rome "a little ere the mightiest Julius fell" (1.1.114). Narrative Insets, therefore, wedge apart the present tense surface of the play and carve corridors of varying depths into the past, opening to our experience levels and recesses of temporality we could never have known if the mode were purely dramatic.

Berry remarks that "all through its course, the action of *Hamlet*—because that action is throttled, its straightforward frontal plot-line is impeded—tends to explode into compensatory Insets."[27] This extraordinary prevalence of Insets in *Hamlet* implies a certain incompleteness of the present that thwarts understanding. The past could give meaning to the present, but the past has been suppressed by the lies of Claudius, whose murder has severed the continuity of time and left Denmark adrift in a discrete present given over to hidden motives and deceptions. The present "this" keeps happening, but it is repeatedly presented as a baffling experience:

> What may this mean . . . ? (1.4.51)
>
> What means your lordship? (3.1.106)
>
> Will they tell us what this show meant? (3.2.153)
>
> Alas, sweet lady, what imports this song? (4.5.27)
>
> What should this mean? . . .
> Or is it some abuse, or no such thing? (4.7.48–49)
>
> O proud Death,
> What feast is toward in thine eternal cell,
> That thou so many princes at a shot
> So bloodily hast struck? (5.2.366–369)

We in the theater are similarly baffled by the "this" of the dramatic mode, and would remain permanently baffled if the present tense immediacy of our experience were not mediated by narrative Insets—most crucially, that of the Ghost telling its story to Hamlet. The Ghost, as the revisiting past, is the patron spirit of narrative Insets. It interrupts and momentarily retards the present in Denmark precisely as the Inset interrupts and retards the dramatic present tense mode in the theater.

But perhaps the greatest retardation comes by means of Hamlet's soliloquies, which, though not narrative, are a kind of Inset. Even more than the episodes of Beckerman and the scenes of Rose, the soliloquies are independent of the plot line. As Harold Fisch says,

> They are much more radically inward-turned, having reference to an inner drama abstracted from the Ghost and the assassination, and centering on Hamlet's search for the meaning of life. They exhaust themselves in the analytic process itself without serving in any obvious way to advance the action. As a sign of this it is notorious that the order of the soliloquies is to some extent interchangeable.[28]

This indifference of the soliloquies to the causal processes of the plot is best illustrated by the third soliloquy, beginning "O, what a rogue and peasant slave am I!" This speech gives an appearance of causal functioning because it works its way toward the practical conclusion, "The play's the thing/ Wherein I'll catch the conscience of the King" (2.2.605–606), which prepares us for the presentation, one scene later, of "The Murder of Gonzago." Yet despite this appearance of arriving at a practical decision by way of soliloquy, the speech is to all functional intents entirely superfluous. Hamlet's decision to catch the conscience of the King with "The Mousetrap" had already been made prior to his soliloquy. Thus as the players were departing earlier, Hamlet detained his "old friend" to ask, "You could, for need, study a speech of some dozen or sixteen lines, which I would set down and insert in it, could you not?" (2.2.540–543). Logically, of course, this remark should not precede but immediately follow Hamlet's soliloquy so that the line about catching the King's conscience would lead directly to its implementation. As it is, the inversion of the two passages renders the soliloquy functionally irrelevant; it could be erased from the play without affecting the plot. And yet in itself it constitutes one of the most brilliant and engrossing moments of the play. It is as though Shakespeare had deliberately inverted the syntax of causation here in order to say, "You see, Hamlet's brief remark to the First Player is all that is strictly necessary for the functioning of the plot. But consider what is lost if we try to live by strict necessity alone. How thin a line mere function draws—and by contrast what depths and shadings and curious passageways of experience lie to hand if, like the imaginative Prince, we but pause a moment."

Hamlet's soliloquies interrupt the action and plot-line of the play even more incisively than the narrative Insets because they are conducted, whatever their literal tenses and moods, in the subjunctive, where everything is conditional. Of course Hamlet himself occupies the present tense indicative zone of drama insofar as he stands before us musing in the here-now. But his musings themselves dwell in the regions of "if," in an indeterminate not-here, not-now, not-there, not-then. *If* this sullied flesh would melt, *if* the Everlasting had not fixed his canon against self-slaughter, *if* this player here who weeps for Hecuba had the motive and the cue for passion that Hamlet has, *if* the dread of something after death did not puzzle the will, *if* Hamlet killed Claudius at his prayers, *if* Hamlet were to find quarrel in a straw like Fortinbras. The narrative there-then steps outside the dramatic present into the past, but the subjunctive steps outside time itself. The imaginary has no tense. Hamlet's soliloquies form a parenthesis in time and a suspension not merely of Hamlet's own actions in Denmark but of Shakespeare's dramatic mode itself, by which his play flickers from one present instant to the next, spending itself to end itself, making haste on its own decay.

Shakespeare's resistance to the kinetic consummations of the dramatic mode is paralleled by Hamlet's resistance to the degenerative processes of the court in Denmark, where his royal father has metamorphosed into Claudius, the funeral has been erased by the wedding, and the chivalric past has succumbed to a present of roistering indulgence as the flagons of Rhenish pass round and the King and Queen post themselves to incestuous sheets. Only Hamlet, motionless and silent when we first see him, grieving backwards, and more than half in love with easeful death, is the anomaly. He is as alien to the dramatic present as the Ghost. Thus in his first soliloquy we find him erecting verbal as well as emotional barriers to the diachronic, wishing to annihilate both himself and the passage of time, yet incapable of doing either. Thus the very style of his thought is characterized by delay in the form of redundance, repetition, and interruption. His first sentence—

> O, that this too too sullied flesh would melt,
> Thaw, and resolve itself into a dew! (1.2.129–130)

—is repetitive and triply redundant. It is as though Shakespeare has dramatized the paradigmatic process of selection that takes place as Hamlet searches for the right expression. The synonymous vertical terms that could substitute for "melt" have been incorporated into the horizontal sequence so that the movement from "melt" to "thaw" to "resolve itself into a dew" foregrounds the synchronic at the expense of the diachronic.[29] A similar effect results from the piling up of adjectives in line 133—"How weary, stale, flat, and unprofitable"—and from the doubling of adjectives in "things rank and gross in nature" (136). In addition to "too too sullied flesh" there are such repetitions as "O God, God," "Fie on 't, ah fie!," "Nay, not so much, not two," "Why she, even she," and "within a month . . . A little month . . . within a month." Hamlet is as reluctant here to get on with the business of syntax as he is spiritually to get on with the business of living.

Interruptiveness is a dominant characteristic of Hamlet's soliloquies.[30] The disjunctive style that results is in radical contrast to the court urbanities of Claudius, with their easy rhetorical paradoxes and sustained periods, and even to the equally troubled but stylistically dissimilar soliloquy of Macbeth beginning "If it were done when 'tis done." Hamlet's style is spasmodic and self-reflexive, pausing and looping back upon itself with comments, exclamations, corrections, and questions that seem to come from somewhere outside the main current of thought. In the first soliloquy, for instance, there are at least seven interruptions: "O God, God . . . Fie on 't, ah fie! . . . Nay, not so much, not two . . . Heaven and earth,/ Must I remember? . . . Let me not think on 't. Frailty, thy name is woman! . . . O God, a beast, that wants discourse of reason,/ Would have mourned longer . . . O, most wicked speed, to post/ With such dexterity to incestuous sheets!" Each of these could have been uttered by a second party, if we substituted appropriate pronouns—say, by an un-stoic Horatio attending to Hamlet's words with the kind of empathic intensity with which Hamlet himself attends to the Ghost's story later ("O God! . . . Murder? . . . O my prophetic soul!" [1.5.25, 27, 41]). By means of these interruptions Hamlet, as though to transcend the solitude of soliloquy, generates within his own utterance a responsive audience. Interior monologue becomes interior dialogue. In the process

Hamlet converts the diachronic subjective utterance, the flow of the ego, into synchronic objectivity by assuming an outside stance, by adding to the text of soliloquy a metatext devoted to self-critical comment and query. Insofar as he creates a division of texts, he creates also a division of selves. The arrests and releases, interruptions and junctures in his speech imply a discontinuous identity shifting from ego to other—from "I" to second-self-"I" who perceives the first "I" as "you," as object.[31]

This element of division and dialogue in Hamlet's soliloquies helps account for our feeling that we are being presented not with the conclusions of thought but with thought itself. For the second voice issues, as it were, from the vertical axis of language, which governs the process of selection. It arrests the continuity of discourse in order to explore the full range of expressive possibilities. The horizontal axis requires a choice to be made—a verb to agree with its subject, an object to take the impact of a transitive verb, an adjective to enhance an adverb, and so on—but the vertical axis prefers to linger over its decision, to dwell awhile in the tasting room of vocabulary and meaning.[32] And instead of acceding automatically to the either/or demands of the diachronic, it sometimes pursues its own policy of both/and. Instead of selecting either "rank" or "gross," let us have "things rank *and* gross in nature." Instead of choosing either "pales" or "forts," let us have "the pales *and* forts of reason" (1.4.28), and so on, thusly:

> Within the book and volume of my brain (1.5.104)
>
> The slings and arrows of outrageous fortune (3.1.59)
>
> The whips and scorns of time (3.1.71)
>
> And enterprises of great pitch and moment (3.1.87)
>
> If his chief good and market of his time (4.4.34)
>
> That capability and god-like reason (4.4.38)
>
> Witness this army of such mass and charge (4.4.47)
>
> That for a fantasy and trick of fame. (4.4.61)

William Empson observes that this doubling of terms "is rare before Shakespeare, and even in Shakespeare before *Hamlet;* it is not likely to be sought for by an author unless he wants to hold a thought in the reader's mind while he plays round its implications. . . ."[33] That is, we can view this doubling of terms—along with various other forms of artistic arrest—as a refusal to relinquish the sovereignty of the moment, its full range of possible being, to the functional demands of time and action. Indeed, so pervasive is this issue in *Hamlet* that we may wonder if Shakespeare is not dramatizing the creative process itself, particularly that insurrectional interim between the first vague motions in the mind and their final clarification as created form—that period in which the playwright rummages through the thesaurus of his imagination, the period of the temptation of his art, when he defers as long as possible the fatal act of choice (in which his freedom to choose destroys his freedom to choose), the inevitable necessity of committing himself to one word, one metaphor, one speech, one scene, one play. That creative act is truly a "consummation," to use Hamlet's term—both a desired achievement and a consuming-away of the alternative options.

However, as if in defiance of this consuming-away by exclusion, Shakespeare seems to have incorporated into the texture and structure of his play that pre-elective moment when the imagination hesitates before a host of possibilities. Merleau-Ponty, talking about painting, tells about an experiment in which that moment just prior to creation was registered by a slow-motion camera focused on the brush of Matisse:

> The impression was overwhelming—to the point that, they say, Matisse himself was moved by it. The same brush which, to the eye, did not jump from one movement to another, could be seen meditating, in a suspended and solemn time, in an imminence, like the beginning of the world, beginning ten possible movements, performing in front of the canvas a sort of propitiatory dance, coming so close several times as almost to touch it, and finally coming down like lightning in the only stroke necessary.[34]

Like the slow-motion camera, Shakespeare records in *Hamlet* not merely the final necessary strokes of signification but the propitiatory dance of the mind as well, thus retaining, as it were, all ten possible movements. A major way of managing this inclusion of the excluded is by means of negation, as discussed in the previous chapter, since negation possesses the remarkable gift of lending presence to absence. A similar effect is created by the kinds of artistic arrest discussed above. As diachronic negations they bring into imagined existence various linear nothings—the pause of Pyrrhus' sword, of Hamlet's voice in soliloquy, of Hamlet's revenge in general—so that the absence of a middle from one point of view becomes, from another, its presence. This act of giving a name to nothing, of filling a diachronic emptiness with the term "zero," is a spatializing of the temporal. The passing moment is arrested and blocked out. "Doing" is placed in suspension so that "being" (or, more accurately, "not-being") may be explored. Liberated from its role as dispensable means between past and future, the present is made momentarily an end in itself.

From this perspective the play is somewhat like a sentence that refuses wholly to erase its provisional ephemeral meanings, those semantic sideroads we glimpse in passing as we negotiate our way toward a final reductive meaning issuing from an Aristotelian entelechy.[35] Or, to adapt a metaphor from the play itself, the meaning of *Hamlet*, or rather the meanings of *Hamlet*, are not to be found entire in the graveyard of its ending. There we encounter only what remains after the flesh has fled. There *is* that meaning, to be sure—the skull of Yorick lies heavy in Hamlet's palm. But there are other meanings as well—gibes, gambols, songs, flashes of merriment, the back on which a small boy rode, memorial moments caught from the flux and cherished in the mind. Yorick means far more than his skull. And the significance of *Hamlet* stands in rich excess to the meaning written on its winding-sheet.

THIRTY-TWO

REWORDING THE MATTER

WE SEEM by this time to have come half-circle. We began this chapter by calling attention to the functional erasure of medial terms in communicative sequences, and we have got round now to noting ways in which Shakespeare not merely rescues medial terms from oblivion but even aggrandizes them to the point where they momentarily arrest and erase the functional process itself. Let us pursue this matter a bit further by focusing on messengers and messages as medial substitutes for visual immediacy. That is, let us consider as "immediate" those situations in which an event is directly witnessed by an observer, as for instance the Ghost—"this dreaded sight, twice seen of us" (1.1.25)—was witnessed by Bernardo and Marcellus. Then, as "mediate," we have those situations in which this direct experience of an event is conveyed to someone else by a messenger-narrator and, more importantly, by his message-narration, as when the two soldiers tell Horatio about their ghostly experiences. Language, however, is not always a convincing substitute for direct experience, and hearing, as Marcellus observes, is not believing:

> Horatio says 'tis but our fantasy,
> And will not let belief take hold of him. (1.1.23–24)

Skeptical stoic that he is, Horatio has come to the castle platform in the cold middle of the night to put the two soldiers' story to the test. At this point we encounter the first narrative Inset of the play, Bernardo telling how—

Last night of all,
When yond same star that's westward from the pole
Had made his course t' illume that part of heaven
Where now it burns, Marcellus and myself,
The bell then beating one—
Enter Ghost
Mar. Peace, break thee off! Look where it comes again! (1.1.35–39)

Here we have a familiar example of functional erasure. The startling appearance of the Ghost erases the conclusion of Bernardo's sentence by substituting visual immediacy for verbal mediation, the phenomenon itself for an account of it. It also erases Horatio's doubts:

How now, Horatio? You tremble and look pale.
Is not this something more than fantasy?
What think you on it?
Hor. Before my God, I might not this believe
Without the sensible and true avouch
Of mine own eyes. (1.1.53–58)

The priority of seeing over hearing is powerfully asserted. Verbal representations of the Ghost's invasion are secondary, parasitic, dubious—and quite dispensable once they have performed their function of ushering Horatio (and later Hamlet) into the visible presence of the Ghost.

A second, rather different case: the Ghost's own story about the murder of Hamlet's father. Here again seeing is believing; for although Hamlet readily accepts the Ghost's claims when he is in the presence of the Ghost itself, he later, left with the story alone, entertains doubts that prompt him to put it to the test. That test, however, is complicated by the fact that the originating event, the murder in the orchard, cannot be retrieved. Hamlet's father can return from the past as a visible, verifiable Ghost, but the occasion of his death requires even ghostlier mediation. It must pass through a series of substitutes: the Ghost's story, "The Murder of Gonzago," and the reworded version of that play which Hamlet calls "The Mousetrap." As purely pragmatic instrument, bait for the kingly conscience, "The Mousetrap" is self-erasing. The King does not really see it, he sees through it to the original events it represents; and Hamlet and Horatio scarcely look at it:

Give him heedfull note,
For I mine eyes will rivet to his face,

> And after we will both our judgments join
> In censure of his seeming. (3.2.83–86)

Where Horatio tested the soldiers' story by reference to the phenomenon itself, Hamlet tests the Ghost's story indirectly, by reference not to the originating event of murder, nor to its pragmatic substitute on stage, but to the receiver of this communication, the King. What Hamlet sees, of course, amply verifies the Ghost's story.

But what if the eye fails entirely, if there is no seeing to create believing? That is the situation in Gertrude's closet when the Ghost reappears, but not to Gertrude. "Do you see nothing there?" Hamlet cries. "Nothing at all; yet all that is I see" (3.4.136–137). And she continues:

> This is the very coinage of your brain.
> This bodiless creation ecstasy
> Is very cunning in.
> *Ham.* Ecstasy?
> My pulse, as yours, doth temperately keep time,
> And makes as healthful music. It is not madness
> That I have uttered. Bring me to the test,
> And I the matter will reword, which madness
> Would gambol from. (3.4.143–151)

Gertrude does not bring Hamlet to his proposed test, but we are obviously to infer that had she done so he would have passed it. But how? By what criteria could Gertrude distinguish Hamlet's rewording of this ghostly matter from his mad retelling of an hallucinatory experience? How could Hamlet make nothing become something for Gertrude? Not, surely, by adopting Ophelia's mad style.

> Her speech is nothing,
> Yet the unshaped use of it doth move
> Her hearers to collection; they yawn at it,
> And botch the words up fit to their own thoughts. (4.5.7–10)

In Ophelia's speech communication disintegrates as words become arresting, not for their astonishing rightness, but for their pure astonishment as verbal Rorschalks inviting total free play from her hearers.

Hamlet, however, makes a remarkable claim for his own words, if brought to the test. For Bernardo and Marcellus the "test" merely involved matching the sign with its visible referent, the Ghost, making the heard-story answer to the seen-fact. But Hamlet has already said "Look there!" and Gertrude, looking, has seen nothing. Far from being the verifying test of his verbal claims, the Ghost is now entirely dependent on Hamlet's mediating language. His rewording of the matter will somehow capture the faded ghostly reality and convey it so truly to Gertrude that she will be persuaded that her dead husband's spirit was indeed present. His words will act upon Gertrude's imagination so as to produce poetic faith in the absence of empirical proof. She will be made to see through the power of speech alone. And at the same time Hamlet will prove that he is not mad like Ophelia, nor a victim of "fantasy," as the soldiers were accused of being, but is "mad in craft" (3.4.195). Although Hamlet is not put to this test, yet throughout this scene, and especially in this speech, his words do confirm the existence of the Ghost for Gertrude—as a symbol of her own "trespass":

> Mother, for love of grace,
> Lay not that flattering unction to your soul
> That not your trespass but my madness speaks. (3.4.151–153)

Thus at the conclusion of his speech Gertrude cries, "O Hamlet, thou hast cleft my heart in twain." Through the instrumentality of Hamlet's eloquence, nothing has become a most potent something.

In the first two examples of mediation above, the message (Bernardo's suspended story and "The Mousetrap") is merely a dispensable substitute for immediate experience. Its value depends on the degree to which it corresponds to the original event, as judged directly by the event itself (the appearance of the Ghost) or indirectly by the effect of the message on one who remembers that event (Claudius). In the third example, Hamlet's rewording of the matter will, like "The Mousetrap," recreate an event that is no longer available for examination. Unlike "The Mousetrap," however, there is no direct or indirect means of verifying Hamlet's account, the truth of which lies not in its correspondence to the Ghost but in its internal consistency and evocative power, in its capacity to bring the Ghost into the closet of Gertrude's imagination. Gertrude can be made to see the Ghost only through the agency of Hamlet's rewordings.

In a metaphoric sense, "seeing the Ghost" by means of rewordings of the matter occurs throughout *Hamlet*. Let me explain this by reference to the journey, for it is a curious feature of the play that it is replete with comings and goings, at least sixteen of the one and eleven of the other.[36] The Ghost, for instance, makes no fewer than five round-trip journeys between its otherworldly prison-house and Elsinore, three of which are partially dramatized. In addition, Hamlet, Horatio, and Rosencrantz and Guildenstern all arrive in Elsinore from Germany; Laertes arrives from France, departs, and comes again; Fortinbras arrives from Norway, departs for Poland, and arrives again from Poland; the players arrive from England; Hamlet and his schoolfellows depart for England, he to return by pirate ship, they to return no more; and various ambassadors pass between Denmark and both Norway and England. Then too there are travels of a nearer sort, such as Hamlet's distracted journey to Ophelia's chamber, and others quite distant in space and time—to the "high and palmy state of Rome,/ A little ere the mightiest Julius fell" (1.1.113–114) and even to far-off windy Ilium on Priam's death-day (2.2.450–518). One effect of these comings and goings is to expand the horizons of Denmark to include France, Germany, Norway, England, Poland, and that "undiscovered country" on the far side of death. A geographical scope is imparted to the world of the play that is analogous to the temporal scope lent it by the exceptional number of narrative Insets, as discussed earlier. These narrative Insets in *Hamlet* are somewhat like the choruses of *Henry* V, which waft Harry's army back and forth across the Channel and enable Shakespeare to shift his scene from England to France. In *Hamlet*, however, there is no shift of scene; Shakespeare strictly observes the unity of place. Thus whereas in *Henry* V the scene itself travels by courtesy of the choruses from here to there, in *Hamlet* the scene in Elsinore remains stationary, and the outside world comes and goes by means of narrative Insets.

At first glance we assume that the narrative Insets are caused by the originating events and places they refer to, that they are verbal substitutes for various happenings in England, Norway, Rome, Troy, Ophelia's chamber. But of course the reverse is actually the case: the war of the theaters in "the city," the drowning of Ophelia, Hamlet's adventures at sea, all are "caused" by the narrative Insets that tell of them. What happened offstage is not re-presented in narrative form, because it does not exist apart from that narrative form. Only the present is "present," a voice speaking, rewording a matter that has no being outside its words, calling up for our imaginative apprehension ghostly occasions that, like Gertrude, we can see only in the mind's eye: two kings in mortal combat, the survivor poisoned years later in his orchard, his son holding Ophelia hard by the wrist and perusing her face as he would draw it, Fortinbras forswearing an assay of arms against Denmark, the child actors berattling the common stages of the city, Pyrrhus' sword making its delayed but fatal descent, and so forth. Whatever takes place offstage, outside the present tense immediacy of the dramatic mode, is but a projection of that mode. The world offstage resides only in the word onstage. It is language we hear and ghosts we imagine.

THIRTY-THREE

THEATER AS GO-BETWEEN

FROM THIS point of view, "seeing the Ghost" is rather like naming Claudio, since without his name Claudio would not exist in Elsinore, and without narrative Insets, which "name" events elsewhere, nothing beyond the immediate here and now of Elsinore would exist in *Hamlet*. Nothing would remain but Elsinore. And yet . . . must give us pause. Is Elsinore really one epistemological step nearer the audience than Wittenberg and Paris and Troy? Well, from one perspective the answer is Yes, since, as argued in the previous section, we directly see (and hear) what happens in Elsinore but can only hear about what happened elsewhere. But from another perspective the answer is No, since what we actually see and hear occurs not in Elsinore but in the Globe theater. That is, just as the narrative Insets brought verbal accounts of, say, Trojan and Norwegian affairs to various hearers in Elsinore, so *Hamlet* brings a dramatic account of affairs in Elsinore to Shakespeare's audience in London.

In other words, the play itself is part of still another communicative sequence. In that sequence the Globe theater (actors, stage, props) is the messenger by which a message (the performance of *Hamlet*) about a place and an event (Elsinore and what happens there) is conveyed from the playwright Shakespeare to his London audience. In this sequence, we should note, two erasures take place. The first is the erasure of the playwright, since in the dramatic mode, as Stephen Dedalus observes, "the artist, like the God of the creation, remains within or behind or beyond or above his handiwork, invisible, refined out of existence, indifferent, paring his fingernails."[37] This erasure of the playwright leaves the audience in the position of someone receiving a written communication, having access only to the messenger and the message. Let us for convenience collapse these two medial terms, the messenger (actors, stage, props) and the message (dramatic performance), into one and refer to it simply as "theater," the theater as go-between. This gives us a large-scale parallel to the narrative Insets, which seemed to *re*-present but actually presented constitutively for the first time the occasions of which they spoke. Here, then, is our second erasure, for just as the narrative Insets erase their past referents by naming them—and just as the word "Claudio" erases that postal functionary by naming him—so the theater erases the apparent existence of Elsinore and its tragic affairs by giving them dramatic expression. The dramatic "middle" erases its "beginning" by usurping its role. Or, even more confusingly, the theater that substitutes for Danish realities actually creates (and therefore erases) the realities for which it substitutes. Elsinore cannot exist without the Globe; but because of the Globe, Elsinore does not exist.

But of course this is true of all imaginative literary works. Why, then, should it seem so special with *Hamlet*? Primarily because Shakespeare insists that we take note of it. It is a critical commonplace that Shakespeare has pervasively theatricalized life in Denmark not merely by staging a fairly full-fledged Inner Play but also by presenting Claudius as an inauthentic player-king and Hamlet as a player-madman, by fashioning various dramatic Insets—for example, the Ghost's arrival performed for an audience on the castle platform, the Nunnery Scene performed by Hamlet and Ophelia for the "lawful espials" Claudius and Polonius, and the sword-"play" performed by Hamlet and Laertes for their courtly audience—and by studding the language of *Hamlet* with such theatrical terms as "act," "show," "play," "audience," "argument," "perform," and so on. In large and small ways these instances of theatricalization in Denmark serve as Brechtian alienation devices to shatter our illusion of Danish reality and cut the cord of our imaginative life there. When Hamlet reviews the First Player's performance of Aeneas telling Dido of the death of Priam and the grief of Hecuba, he says—

> Is it not monstrous that this player here,
> But in a fiction, in a dream of passion,

> Could force his soul so to his own conceit
> That from her working all his visage wanned,
> Tears in his eyes, distraction in his aspect,
> A broken voice, and his whole function suiting
> With forms to his conceit? And all for nothing!
> For Hecuba!
> What's Hecuba to him, or he to Hecuba,
> That he should weep for her? (2.2.551–560)

Hamlet's separation here of the First Player's performance from the apparent experience of Aeneas renders it all merely "a fiction . . . a dream of passion." But it also obliges us to separate Richard Burbage's performance from the apparent experience of Hamlet, and thus to regard the anguish of the soliloquy itself as merely "a fiction . . . a dream of passion." It is not Hamlet we have been watching, the play tells us, but Burbage, not Ophelia but a boy actor, one of the "little eyases" of Shakespeare's company. And as for life in Denmark with its marvelous intensities of feeling, its flashes of eloquence, and its absorbing events? Merely a dream on stage, a fiction, a play. Alas, that it should come to this! Greasepaint and costumes, actors with memorized speeches and tears teased into their eyes. And all for nothing! For Hamlet? What is Hamlet to Burbage, or he to Hamlet, that he should weep for him?

We must, it seems, ask these questions, thus doing to Hamlet what he does to Aeneas. Otherwise we shall find ourselves unhappily associated with Polonius, who in his university days enacted Julius Caesar, and so threw himself into his part that even now he cannot distinguish himself from it: "I was killed in the Capitol; Brutus killed me" (3.2.101–102). This same inability to distinguish the actor from his role causes Polonius to sentimentalize the performance of the First Player as Aeneas and call it abruptly to a halt: "Look whe'er he has not turned his color and has tears in 's eyes. Prithee, no more" (2.2.519–520). Hamlet is more sophisticated—and who would not prefer Hamlet's account of matters to that of Polonius?

A small question. The large question is, Why should Shakespeare want to sabotage the illusion of reality in his play? Perhaps a piece of an answer will be forthcoming if we key on the Ghost once again. As we have seen, Shakespeare goes to considerable pains to render the Ghost credible in the opening scene. Even the skeptical Horatio, who represents the skeptic in us, is obliged to let belief take hold of him. And Hamlet, when he comes into the awesome presence of the Ghost, is similarly persuaded. And so are we. Our willingness to suspend our disbelief in the Ghost represents more largely our poetic faith in the reality of Horatio and Hamlet and the tragic affairs of Elsinore. We believe the Ghost, we believe *in* the Ghost, and we believe in life and death in Denmark. But after the Ghost has disappeared, leaving Hamlet with his anxious companions, Hamlet adopts a fooling mood and, as Bridget Gellert Lyons says, when the Ghost cries "Swear" he "attempts to reduce the Ghost to a stage figure by his reference to it (in terms of the actual stage) as the 'fellow in the cellarage' (1.5.152)."[38] The repeated cries of the Ghost from below the stage during the remainder of the scene, and Hamlet's gamesome comments about it—"Well said, old mole! Canst work i' the earth so fast?" (1.5.163)—literally undermine our poetic faith in the Ghost and return us abruptly from Elsinore to the Globe theater. At a curiously crucial point in the play, it seems, Shakespeare erases the illusion of reality in Denmark and presents us instead with the material inadequacies of the stage. All that we have seen and accepted up to this point seems suddenly fraudulent.

And why should Shakespeare take such risks? In part because he wants our disillusionment to mirror Hamlet's. Hamlet's "problem," after all, centers in the decentering of his world view. As Ophelia informs us, Hamlet's was a "noble mind" that has been "o'erthrown":

> The courtier's, soldier's, scholar's, eye, tongue, sword,
> The expectancy and rose of the fair state,
> The glass of fashion and the mold of form,
> The observed of all observers, quite, quite down! (3.1.153–156)

He who was once the glittering embodiment of Renaissance values, a princely courtier straight out of Castiglione, hears the Ghost's story of murder and betrayal, and his response is to obliterate the Renaissance world picture on which his own being is founded:

> Yea, from the table of my memory
> I'll wipe away all trivial fond records,
> All saws of books, all forms, all pressures past
> That youth and observation copied there,
> And thy commandment all alone shall live
> Within the book and volume of my brain,
> Unmixed with baser matter. (1.5.99–105)

For Hamlet the Ghost's story dissolves, annuls, or in the fashionable term deconstructs his schoolboy conceptions of a world governed by honor, love, and truth. Its cosmetic surfaces scraped away by the Ghost, life in the court of Claudius—and, to Hamlet's generalizing mind, life everywhere—festers with evils as repellent as the lazarlike skin of his poisoned and dying father. As self-appointed pathologist he takes it upon himself to chart the nature and extent of the disease, in both the court and his own psyche, before essaying the Ghost's remedy of revenge.

So Hamlet discovers behind the arras of his youthful assumptions the fraudulence of the world, and suffers for his knowledge. And how are we, as audience, to register his disillusionment? In some degree we must, as we would in any play, enter the hero's world imaginatively and see it from within. We must be caught up in the unforgettable intensity of his tragic milieu and register through his consciousness a world in which the bottom has fallen out of all values, feeling with him how weary, stale, flat, and unprofitable it all is. But this empathic engagement with Hamlet, this passage out of our lives into his life in Denmark, is too easy—any Partridge or Polonius can do it. And Shakespeare is not content for us to do our living by means of Hamlet. He insists that our passive identification with the hero and acceptance of his world yield to, or at least be supplemented by, an active involvement within our own sphere of experience. Thus by theatricalizing the Ghost, Shakespeare says to us, rather as Hamlet says to his mother, "Not this, by no means, that I have bid you see." As a result, our dramatic faith in the illusion of real life in Elsinore collapses, precisely as Hamlet's belief in an idealized world of goodness, beauty, and truth collapses when he hears the Ghost's story. In the theater, as in the court, all is seeming.

But of course we have never really believed otherwise. Or so, at least, Doctor Johnson claims. In his admirable refutation of the need for the unity of time, Johnson dismisses the notion of mimetic presence in drama—that is, of an apparently real world populated by real people with whom we become emotionally engaged. With the air of a man refuting Bishop Berkeley by kicking a stone, he says, "The truth is, that the spectators are always in their senses, and know, from the first act to the last, that the stage is only a stage, and that the players are only players."[39] That is well and decisively put. But one wonders if this Theseus-like advocate of "cool reason" is the same Doctor Johnson who writes in his notes on *King Lear*,

> And, if my sensations could add any thing to the general suffrage, I might relate, that I was many years ago so shocked by Cordelia's death, that I know not whether I ever endured to read again the last scenes of the play till I undertook to revise them as an editor.[40]

Or if it is the same Doctor Johnson who says of Hamlet's speech over the praying Claudius that it "is too horrible to be read or to be uttered" and that the appearance of the Ghost "chills the blood with horror." The editor writing critical prefaces to Shakespeare in his study is evidently quite another person from the theater-goer directly experiencing *King Lear* and *Hamlet*.

Nevertheless, Johnson is right—but also wrong. On the one hand, theatrical art is more honest about its mediating role, about the presence of its signifiers, than purely verbal works, in which language lays the sly film of familiarity over its signifiers and induces us to feel that we are in the presence of the signified itself. Language hides what theater advertises: to read a book one does not have to check the newspapers for opening dates and times, hire a cab, buy a ticket, enter a building and find a seat, listen to incidental music, and wait for the foot-shuffling and coughing to subside. It is hard for theater to erase its own presence. "Theater," Eric Bentley observes, "is shamelessly 'low'; it cannot look down on the body, because it *is* the body."[41] Whereas language possesses meaning but lacks substance, plays, by virtue of their embodiment in actors on stage, are gifted with a substantial presence that words in poetry and prose can achieve only by deliberately foregrounding the physical properties of sound and rhythm.[42] This corporeality of the theater would seem to prevent it from becoming transparent and hence to prevent us from forgetting that we are in the presence of art's signifiers, not of reality's signifieds.

And yet, on the other hand, the relationship between signifier and signified in drama is less arbitrary than it is in language. The actor actually does not "signify" a character so much as he incarnates it. Richard Burbage does not arbitrarily stand for Hamlet; he intentionally paints Hamlet with his body and voice. There is always between actor and character a resemblance. Even if Hamlet were played by a puppet, the aim would be to achieve a puppet-like Hamlet or a Hamlet-like puppet, to surprise and please the audience with the resemblance, however distant. Thus the space or deferment between actor and character, and between stage and the depicted scene, is less great than it is between verbal signifiers and their signifieds; and thus we are more readily persuaded that we are seeing not Burbage but Hamlet, and even not a boy actor but Ophelia.

Moreover, the elaborate preparations required for us to come into the presence of the dramatic work—all the ticket-buying and seat-taking and the minor physical adjustments to the theater and other theatergoers—prepare us also to accept the conventions of our context. We have come here, that is, not to see Richard Burbage and the stage of the Globe but to see Burbage-as-Hamlet and the stage-as-Elsinore. We shall not allow the mediations of theater to become a get-between but will transform them into a go-between. For in this theatrical context, it is not an artificial but a perfectly natural act for us to enter the "as-if-ness" of fictional life, whatever might be natural in other contexts. The commonsense of Doctor Johnson in his study and the rational skepticism of Duke Theseus about the products of the imagination are neither sensible nor rational in the theater, where it is right to imagine and wrong to reason, where "seeing things" is a sign of sanity, and ghosts are invisible only to Gertrude.[43]

For these and other reasons, we fully expect and are more than willing to "see the Ghost" of fictional Elsinore that Shakespeare evokes for us. And yet, as I mentioned, Shakespeare evokes the Ghost with one hand and erases it with the other: "Not this, by no means, that I have bid you see." With that, we are returned from the Elsinore of our imaginings to the cold hillside of theatrical fact, to what Shakespeare called in *Henry* V the "mockeries" of the stage. Perhaps it is no accident that in his soliloquy of erasure after hearing the Ghost's story, Hamlet speaks of "this distracted globe," for it is not merely his mind but the Globe itself, and the minds of the audience within it, that are distracted.

And to what end? If Shakespeare is intent on disillusioning us, as the Ghost does Hamlet, and if he succeeds, does he not then ruin the play for us? Before we answer that question, we should answer another: Does the Ghost's disillusionment of Hamlet ruin life for him? The answer is, Indeed it does, so much so that he can scarcely bear to continue living. But he does. When Ophelia's world collapses from the loss of lover and father, she sinks into madness and a watery grave. Hamlet, however, does not go mad when his illusions dissolve but plays mad, and he does not die though he rejects life. He exchanges a naive vision of life as devoid of evil, corruption, and death for a cynical vision of life as no more than evil, corruption, and death. For him Denmark is a prison, man a quintessence of dust, Ophelia a breeder of sinners, the King a cutpurse of the empire, his mother an example of reason pandering will, his schoolfellows a pair of adders fanged, and even imperial Caesar now merely clay to stop a hole against the wind. And yet in a blighted world that mocks at all he once cherished, and forces him to mock with it, Hamlet manages not merely to survive but to fulfill a trust. If he once exchanged schoolboy innocence for cynical disillusion, he gradually exchanges that for a more embracing conception of the human condition, one that neither endorses nor denies either innocence or disillusion but acknowledges both. This stoic resignation is most movingly expressed at the conclusion of the "readiness is all" speech when Hamlet says simply, but with the full weight of meaning imparted by the play to that smallest and yet most capacious of verbs, "Let be."[44]

I think we feel this quiescence in Hamlet the better for our having shared in the theater something of the turbulence of his experience in Elsinore. And if hero and audience are alike in their losses, perhaps they are alike also in their gains. As Hamlet senses toward the end a providential design governing the apparently causeless reversals and baffling ironies of human affairs, so we should sense a metadramatic design governing the paradoxes of our theatrical experience. When Shakespeare deconstructs his drama of apparently unmediated signifieds and mimetic presence, we are left confronting nothing but the empty signifiers of theater. Yet the very poverty of the theater calls forth our astonishment, that this bare wooden O and these costumed players could so act upon our imaginations as to evoke a castle in Elsinore and a Danish prince and a Ghost that is "something more than fantasy" not only for Horatio but even for Doctor Johnson himself, his blood chilled with horror. For we have, with Horatio, the "sensible and true avouch" of our own eyes that this place, this prince, and this ghostly thing have come before us. What we have seen, we cannot on command unsee. Commonsense cannot cheat so well as she is famed to do. Nor, so miraculously natural are the workings of imagination, can we withhold our consent to see again, despite Shakespeare's reminders that all is but mockery and theatrical seeming. In these reminders there is construction even in the midst of deconstruction, since the playwright's "Not this, by no means, that I have bid you see" shares with his other negations throughout the play the disconcerting power to affirm while denying.[45]

Perhaps then we realize that like the reversible images of the psychologists—like Jastrow's famous duck-rabbit, for instance—*Hamlet* ultimately draws our attention not to the duck of realistic Denmark or to the rabbit of the Globe theater but to the wonderful coexistence of both in a single form. As a visual pun, they invade one another's territory. The theater brings the castle to the stage by conveying the stage to the castle. That, I take it, is one reason why Shakespeare uncharacteristically introduces into his play the topical discussion of the "little eyases" and the War of the Theaters in which his own company was at this time involved. For by calling the traveling players "the tragedians of the city" instead of, say, "the tragedians of Copenhagen," he invites us to regard "the city" as London and the players as members of the Chamberlain's Company, as of course they are. Those members of Shakespeare's company who play the city-actors are for the duration of the performance on tour in fictional Denmark, in flight, as it were, from their competitive difficulties in London, which are so pointedly associated with the very theater in which they are now performing:

Ham. Do the boys carry it away?
Ros. Ay, that they do, my lord—Hercules and his load too.[46] (2.2.360–362)

The members of the Chamberlain's Men make the journey from London to Elsinore in the time it takes Hamlet to look round and say, "You are welcome, masters; welcome all" (2.2.421). But at the same time that they withdraw from London and arrive as touring actors in Elsinore, they transport the touring actors in Elsinore to London and onto the stage of the Globe theater, where by entertaining the audience they will strike a histrionic blow against the private theaters and the child actors from whose competiton they have fled.

By such ingenious indirections Shakespeare forces us to acknowledge the marvelous and perplexing doubleness that he has worked into the very grain of *Hamlet*. In the duck-rabbit sense, *Hamlet* has become a visual pun writ large. And just as Hamlet's own puns seemed at first divisive, an erasure of primary in favor of secondary meanings, but proved ultimately conjunctive, a circulation of many meanings within the body of one sound, so Shakespeare's metadramatic "Not this, by no means, that I have bid you see" seems at first to divide *Hamlet* into two discrete realms of Elsinore and the Globe—and even to erase Elsinore in favor of the Globe—but then can be seen to set up an indivisible circularity between the two. We are obliged to witness a form of reciprocal possession in which the stage is possessed by the fictions that it itself possesses. For three hours Hamlet must live in Burbage and he in Hamlet, and for that same time Ophelia and Gertrude can speak only through the throats of boys, and the tears that come to the eyes of the First Player in Elsinore run down the cheeks of an actor in London.

Yet despite this indivisibility, despite the fact that we see Hamlet-in-Burbage and Elsinore-in-the-Globe, what we are *not* seeing, the play keeps informing us, is either Hamlet or Elsinore. These we shall never see in their own ideal being, since they exist only in their manifestations. They are like "Claudio," who has no existence apart from his name. Or they are like the dead King Hamlet, who cannot appear in Denmark in his own person but only as "this dreaded sight" whose appearance announces "I am and am not King Hamlet." In this regard the characters and the tragic doings of Denmark are all ghosts returned from an undiscovered country in the form of the actors who play them and the theater they temporarily inhabit. And like the Ghost, who is poised between otherworld and thisworld, so this particular "Hamlet" we see is poised somewhere between Hamlet and Burbage (or Betterton, Garrick, Kean, Gielgud, or Olivier); and the entire play, or this version of it, is poised somewhere between the Danish otherworld and the theatrical thisworld, doomed for a certain term to walk the stage until the cock crows and the curtain falls and Hamlet and Burbage, dispossessed of one another, go their separate ways. In the spirit of "To be and not to be," Shakespeare insists on having it both ways at once, and so requires us in our fashion to say with Hamlet, "Let be."[47]

THIRTY-FOUR

"What Warlike Noise Is This?"

LET ME sort things out a bit before going on. We began with a discussion of functional erasure, especially as regards middlemen in communicative sequences, and noted that Horatio, Hamlet, and the play itself risk self-extinction as in their various ways they become functionaries. At the same time, however, we have seen how Shakespeare resists functional erasure by naming, personalizing, or substantializing his characters or aspects of his play—how the delaying hero individualizes himself at the temporary cost of his murderous mission, how not-revenge and certain other caesurae in the play are endowed with verbal and imagistic presence despite their material absence, how episodic actions and tableaux-like scenes renounce their causative function in the plot and become semi-autonomous, how the narrative Insets and soliloquies retard the diachronic flow of action by abandoning the present-tense indicative-mood of drama, and finally how the theatrical dimension of the play, not content to remain a conventionally self-effacing middleman between Denmark and Shakespeare's audience, foregrounds itself as originative and constitutive rather than mimetically reflective of its tragic fiction.

In all of these cases Shakespeare's style tends toward the discontinuous, digressive, and parenthetical. Which is all well and good as a reaction against the dehumanizing effects of purely functional progression. For to give oneself wholly to such "progress," which in one light is but a quickening toward the consummations of the grave, is surely a form of madness. And yet, as Harold Toliver has shrewdly observed, the digressive and parenthetical, as imaginative deviations from rational sequence, are themselves at the borders of another kind of madness, to which Swift's "Digression on Madness" in A *Tale of a Tub* and Sterne's meanderings in *Tristram Shandy* pay tribute.[48] On this view it is appropriate that in *Hamlet* the hero's digressions from the straightforward performance of his assigned task are denominated "madness," even though only of the "north-north-west" variety. For in Hamlet's obsession with the preconditions of action and the labyrinthine ways of his own being, and in Shakespeare's explorations of the dramaturgy of delay, the digressive and parenthetical risk becoming ends in themselves. When the senile Polonius wanders toward his diagnosis of Hamlet as mad, for instance—

> My liege, and madam, to expostulate
> What majesty should be, what duty is,
> Why day is day, night night, and time is time,
> Were nothing but to waste night, day, and time.
> Therefore, since brevity is the soul of wit,
> And tediousness the limbs and outward flourishes,
> I will be brief. Your noble son is mad.
> Mad call I it, for, to define true madness,
> What is it but to be nothing else but mad? (2.2.86–94)

—our amusement at this tediousness in praise of brevity is tempered slightly by our realization that the old man's balmy disgressiveness is a comic illustration of Hamlet's own behavior as he forgoes the practical business of his revenge to "waste night, day, and time" in addressing himself to such imponderables as "Why day is day, night night, and time is time." Or, as Hamlet himself puts the question,

> The time is out of joint. O cursed spite,
> That ever I was born to set it right. (1.5.189–190)

The comparison is exaggerated, of course, and yet it is true that Hamlet converts a reasonably simple problem into one of vast complexity by making the task of revenge an occasion to explore the nature of human existence, though the exploration itself is infinitely absorbing. Thus as the Queen urges Polonius, "More matter, with less art" (2.2.95), so the Ghost on his return urges Hamlet to more action with less delay. To pursue the parallel: at the point at which Polonius seems about to drift quite beyond all "matter," he nevertheless manages, with the aid of Gertrude's promptings, to fetch about and find his topic—Hamlet's madness; and at the point at which Hamlet seems to have lost all opportunity to perform his task—that is, when, having killed Polonius, he is shipped to England as "this mad young man" (4.1.19)—he too fetches about with the aid of the pirate ship and finds his route to revenge, which is presumably on a heading of south-south-east.[49]

So Hamlet returns from his mad digressions into self to keep his promise to the Ghost, and Shakespeare's *Hamlet* emerges from its long parenthesis of delay to honor its commitment to its revenge genre. But after that violent end comes another end. As Hamlet's world dies into silence we hear off-stage the rattling of drums and firing of cannon, and in steps Fortinbras fresh from his victories in Poland. What are we to make of Fortinbras? We have been led to expect the final revenge as an act of structural fulfillment, but we have not been led to expect the appearance of Fortinbras, especially as Denmark's future king. And what bearing does his entrance have on Hamlet? We have seen that Hamlet escapes becoming a mere functionary for his father by first killing the King in his own behalf. But now it appears that he has killed the King less for himself or for his father than for Fortinbras, that he has swept the way for the accession of the Norwegian prince, becoming even as he dies merely a "voice" cast for the new successor.

At first glance the mysterious Fortinbras seems quite unrelated to both the Danish succession and the tragic experience that ends where he begins. "Who is this Norwegian prince?" Jan Kott asks:

> We do not know. Shakespeare does not tell us. What does he represent? Blind fate, absurdity of the world, or victory of justice? Shakespearean scholars have made a case for all these interpretations in turn. The producer has to decide. Fortinbras is a young, strong and cheerful fellow. On his arrival he delivers a speech to this effect: "Take away these corpses. Hamlet was a good boy, but he is dead. Now I shall be your king. I have just remembered that I happen to have certain rights to this crown." Then he smiles and is very pleased with himself.[50]

On the other hand, Francis Fergusson says, "Fortinbras appears at last in Denmark: a new hope for a new, purged state."[51] Is Fortinbras a new hope or Kott's rash intruding fool? It is hard to say. One thing we can say about Fortinbras, however, is that he is an analogue to Hamlet. Like Hamlet, he shares his father's name, falls into political limbo when his father is killed, sets out to take revenge on Claudius, and discovers by indirections a personal identity in which the claims of both self and son receive their due.

But perhaps Fortinbras is more than merely an analogue to Hamlet. It is a rather curious fact that while Hamlet is alive he appears on stage with every character in the play but one—Fortinbras. Of course Fortinbras only appears on stage twice, at the opening of Act 4, Scene 4 and at the end of Act 5, Scene 2. Yet these two appearances are suggestively similar. On the first occasion Shakespeare offers us a glimpse of Fortinbras pausing, as his army marches across stage toward Poland, and delivering seven lines of characterless instruction to one of his captains. Fortinbras pauses, not for soliloquy or meditation or wordplay, but for business, to get permission to march on. At this point Hamlet and the others enter. But not before Fortinbras has made his exit, taking with him a fine opportunity, one would think, for Shakespeare to dramatize a meeting between the honor-bound man of action and the momentarily envious, dilatory man of thought, the one in command of an army and his own destiny, the other under restraint as a madman, en route to apparent death. It is as though the approach of Hamlet necessitates the withdrawal of Fortinbras, who, rather like Conrad's "secret sharer," rises to the dramatic surface only briefly and cannot appear in the presence of his other self.

The same sort of thing happens in reverse at the final nonencounter of the two at the end of the play. This time it is Hamlet who is on stage and Fortinbras who approaches, heralded by the sounds of off-stage marching and cannon firing. "What warlike noise is this?" the wounded Hamlet asks, and is told by Osric that it is "Young Fortinbras, with conquest come from Poland." Hamlet turns immediately to Horatio:

> O, I die, Horatio!
> The potent poison quite o'ercrows my spirit.
> I cannot live to hear the news from England,
> But I do prophesy the election lights
> On Fortinbras. He has my dying voice.
> So tell him, with the occurrents more and less
> Which have solicited—the rest is silence. (5.2.354–360)

So Hamlet makes his final exit into silence just as Fortinbras makes his noisy entrance into kingship. However, if "silence" here is a metaphor for death, then the gift of Hamlet's "voice" to Fortinbras implies that the dying prince in some sense lives on in the new king.[52]

There is some suggestion, then, that Fortinbras may be viewed not only as Hamlet's chosen political surrogate but also as his substitute self.[53] A severely diminished self, to be sure—merely a shadowy far side of the multifaceted Dane, a synecdochic part left behind to represent the lost whole. But even so, the suggestion that Fortinbras plays the role of alter ego to Hamlet would mitigate somewhat the apparent wrongness of his entry into the Danish succession.

Let me reinforce this view by glancing again at that succession. Had it proceeded normally, "Hamlet" would have succeeded "Hamlet," which is to say that the fictional continuity of "the king's body politic," as Ernst H. Kantorowicz describes it, would have become a fictional circularity underwritten by the genetic, generic, and nominal consubstantiality of the two Hamlets.[54] The intrusion of Claudius, however, both interrupts the succession and wedges apart the shared identity of "Hamlet." The dispossessed Prince finds himself divided between his roles as self and as son, a division that is accentuated during the middle of the play and then is bridged significantly at the end. From the standpoint of Fortinbras, however, the real point of Claudius' intrusive reign is that it makes possible his own kingship in Denmark. If we regard the succession as a communicative sequence in which the crown is the "message," then Claudius, who maneuvered to receive that message from Old Hamlet, now becomes merely a messenger whose function is to pass the crown on to Fortinbras. And like most of the messengers we have examined, Claudius disappears in his function, as we shall momentarily see. But the more immediate question is, How meaningful can the royal message be if it is conveyed to an outsider like Fortinbras?

Perhaps some meaning is restored to this sequence if we observe that the structure of the Danish succession is exactly paralleled in Norway, where Old Norway has popped in between the two Fortinbrases. As a result, certain syntactic equivalences (indicated by "=" in the chart below) emerge between the two successions to complement those "vertically" present within each succession itself:

┌──→ Old Hamlet ←── = ──→ Old Fortinbras ←──┐
= (Claudius) ←── = ──→ (Old Norway) =
└──→ Young Hamlet ←── = – → Young Fortinbras ←──┘

In both countries the descent of the crown from father to patronymic son has been deferred by a horizontal sidestep from brother to brother. The two interim kings are parenthetical suspensions within the syntax of the true succession.

From this standpoint, the accession of Fortinbras, rather than subverting the legitimate succession in Denmark, completes it symbolically in its original form. That is, the original equivalence of the two Hamlets, which seemed superseded by Claudius, is reinstated by the substitution of the Norwegian heir apparent for the Danish heir apparent. The Danish succession is returned, as it were, to the sequence "Old Hamlet–Young Hamlet (=Fortinbras)," a sequence that was not a transfer of kingship but a tautological perpetuation of the reign of the elder Hamlet. The interruptive rule of Claudius thus takes the succession forward to Fortinbras only thereby to take it backward to Old Hamlet.[55] This symbolic restoration of the reign of the dead king is implied visually when the armed Fortinbras makes an entrance at the end of the play that is strongly reminiscent of that made at its beginning by the armed Ghost.[56]

These aspects of the parallel successions endow Fortinbras' "rights of memory in this kingdom" (5.2.391) with a certain legitimacy, thereby meliorating somewhat the pervasive sense of tragic loss at the end of the play. The sense of loss is dominant, of course, and may even seem intensified by the entrance of the Norwegian prince, who is sufficiently unknown, uninformed, and soldierly simple to imply an ironic disparity between the brave new world he seems to represent and the stricken world he comes upon. For Fortinbras—as for Edgar, Malcolm, and Octavius Caesar—it is an unfortunate fact that the tragic experience diminishes all who survive it, those who "shall never see so much, nor live so long," as Edgar says.[57] However, if Fortinbras suffers his part in this general shrinkage of the post-tragic world, at the same time his bluff innocence and incomprehension, which mark him off from the tragic experience, also proclaim his freedom from the perniciousness of Claudian Denmark and suggest that he may indeed be "a new hope for a new, purged state."

THIRTY-FIVE

"To Tell My Story"

THAT FORTINBRAS should constitute both a turning toward the future and a returning to the past, a simultaneous going and staying, is in keeping with the fact that Shakespeare repeatedly dramatizes in *Hamlet* the tension obtaining between the synchronic and diachronic impulses in drama, especially in the form of the pause that conserves and the onward act that expends or destroys. What is most in danger of destruction at the end of the play is the human voice, which is liable, as Hamlet's last words assert, to disappear in silence. Any play, as it makes haste on its own decay, disappears into a vacuum of silence, but *Hamlet* is particularly vulnerable in this respect because the crucial events of its plot keep falling into a well of secrecy. The murder of Old Hamlet exists in silence until it materializes in the form of the Ghost's story. But then it returns to silence again when Hamlet swears his friends to secrecy. The significance of "The Mousetrap"—what it has revealed about the Ghost and Claudius—is relegated to secrecy. And the events of Hamlet's ocean voyage are also lost to silence. Thus when Hamlet says "The rest is silence," the "rest" is not merely the remainder of his life but also the leftover unexpressed occurrences in Denmark that are about to fall into permanent silence. Indeed, silence and secrecy, in their capacity to negate experience in Denmark, become almost positive forces threatening to consume the play entirely, as they seem to do when the last word is spoken—"Go, bid the soldiers shoot"—and the last peal of ordnance faded.[58]

But these various events and voices, though lost in Denmark, are not lost in the Globe theater, in large part because Horatio, in his role as Hamlet's confidant, has delivered them to the Shakespearean audience. Horatio has proved the perfect messenger, keeping Hamlet's secrets silent in the castle, yet conveying them efficiently to the Globe. In fact, so well has he performed this theatrical function that he has virtually disappeared as a real person in Denmark. Reduced to monosyllabic replies during the Graveyard Scene, as we have seen, Horatio himself appears to stand at the grave-edge of silence. Thus it is fitting that when Hamlet is approaching his terminal silence Horatio should claim his own antique Roman right to silence by suicide. However, for Horatio to lapse into total silence would be to violate the meaning of his own name. To fulfill that meaning he must remain alive to minister to the "wounded name" of his friend by giving voice to his story.

Thus Hamlet achieves a certain victory over the silence of death by means of both Fortinbras and Horatio. Fortinbras, the soldier who embodies the principle of functional efficiency, receives Hamlet's "dying voice" and speaks for the executive side of the Prince, for him who was likely, "had he been put on,/ To have proved most royal" (5.2.399–400). That part of Hamlet that survives in Fortinbras is but a diminished version of the Hamlet we have seen throughout the play, he of the preparatory brushstrokes and dark broodings. But perhaps what is lost of this Hamlet in Fortinbras can be recovered by Horatio, who will attempt to square Hamlet's earthly account, not by actions like revenge, but as befits a stoic philosopher, by thought and speech.

Horatio's story will speak not merely for the Hamlet who has died in Denmark but for the play *Hamlet* that is reaching its "consummation" in the theater. The play wards off its own approaching annihilation, by casting the imaginations of its audience forward into a future scene in which Horatio will address his story to the Danish nobles. Thus as the audience in the Globe rises and files out of the theater, another audience, we are to imagine, gathers in Elsinore to hear . . . the story of Hamlet, the first words of which might well be:

> It was bitter cold when Marcellus, myself, and Bernardo,
> The bell then beating one . . . .

The seemingly irreversible march of the play toward its own annihilation has become, we now see, a march toward its recovery and revivification in the story told by Horatio. The mortal linearity of dramatic action will be bent by Horatio's voice into an eternal circle symbolizing the perpetuation of Hamlet's tragedy both in the memories of Shakespeare's immediate audience and in the theaters of the future.

This perpetuation, it goes without saying, is by no means a matter of duplication. As each theatergoer's memory retains a different *Hamlet*, so each new performance by the same company—not to mention each new production—creates a different *Hamlet*. We tend perhaps to think of these performances as different imperfect versions of a platonically ideal *Hamlet*, as various actual circles are imperfect representations of ideal circularity. However, although perfect circularity cannot be perfectly represented in actual circles, it can be geometrically defined to everyone's satisfaction, which is hardly the case with *Hamlet*. Even Horatio's story, which symbolizes the recovery and perpetuation of the play, transforms drama into narrative, and thus loses at least as much in form as it retrieves in content. But perhaps this is to consider the matter too curiously: Shakespeare can hardly ask Horatio to enlighten the Danish nobles and the Norwegian prince by staging a full-scale performance of *Hamlet*, especially since the cast has suffered considerable attrition by the end. Yet even if we made an issue of the transformation of drama into narrative—that is, if we argued that because of this Shakespeare is not implying a circular return to the beginning, as in *Finnegan's Wake*, but a degenerative linearity—a case could still be made for a kind of cyclicism. For as *Hamlet* passes from drama into narrative it returns to its origins in the narratives of Saxo Grammaticus and Belleforest, Horatio taking on the role of latterday historian. Something of this sort is suggested, we might argue, by a similar sequence within the play itself, inasmuch as the Ghost's narrative is converted into dramatic form in "The Mousetrap," which is converted into narrative form again by Horatio. But these arguments seem overly complicated. More simply, the tone and mood of the ending of the play establish Horatio as an authority, an ideal teller who, even in narrative form and despite the obvious limitations of his knowledge, can recapture to everyone's satisfaction all that has taken place. I think, that is, that we must take his word on faith when he says, "All this can I/ Truly deliver."

THIRTY–SIX

A Breathing Space

LET ME conclude this chapter by examining Shakespeare's play on the word "breath," which reflects in smaller compass some of the major issues dealt with earlier. Shakespeare often uses "breath" to refer both to the spoken word and to the breath of life that is expended in the act of speech. Gertrude supplies a perfect illustration of this dual meaning at the end of the Closet Scene when she tells Hamlet,

> Be thou assured, if words be made of breath,
> And breath of life, I have no life to breathe
> What thou hast said to me. (3.4.204–206)

On this view, however eloquent his exhalations, a speaker breathes forth not merely words but also his own life and thus hastens toward the final breathlessness of silence. In a less mortal sense Shakespeare's metaphor underscores the vocal fact that spoken words wither into meaningless air even as they come into being. It is a metaphor that reminds us of the stress in the play on performed drama as a self-destructive achievement, a "consummation." Thus it is in part this verbal ephemerality that the play has resisted in its efforts to give pause to temporal process and bring about states of artistic arrest.

Breath, however, need not always be a death-dealing expense of spirit. It may serve the interests of life. In fact, as "juncture," breath may constitute a pause in temporal process itself, for juncture, as Geoffrey Hartman says, "is simply a space, a breathing space" within an utterance.[59] This is complexly illustrated when the dying Hamlet asks Horatio to absent himself from felicity awhile—

> And in this harsh world draw thy breath in pain
> To tell my story. (5.2.350–351)

Hamlet not merely speaks of juncture in asking Horatio to draw his breath in pain but also illustrates it in his own utterance. For the major breathing space in this passage follows the word "pain," so that at that point Hamlet must draw his own breath in pain. This moment of hesitation is ambiguous both as painful breath and as verbal suspension. For as physiological pause it simultaneously defers Hamlet's death and prefigures it, and as verbal pause it defers the completion of his sentence and at the same time enables his speech to resume.

A similar situation occurs when Hamlet issues his famous phrase "To be, or not to be." The breath he takes where the comma appears, after "be," is at once a prolongation of "being," a hesitation before the ultimate breathlessness of "not-being," and a revivifying pause that allows him to breathe forth the remainder of the line. As this example would suggest, juncture is a pause in speech in which what has just been said is retained in suspension and what remains to be said exists only in a state of virtual being as the brush of the mind hovers silent and contemplative over its canvas. The moment is an imaginative parenthesis between what has been and what will be, an *epochē* of unrealization in which everything is possible. It would appear to be the smallest manifestation of the divisions, pauses, and arrests that we have examined throughout the play, a miniaturization of that period between the command to revenge and the performance of revenge, when Hamlet drew his breath in sustained pain.[60]

But now it is Horatio who is to draw his breath in pain. Of his story Fortinbras says, "Let us haste to hear it,/ And call the noblest to the audience" (5.2.388–389). When Horatio begins to speak, however, all haste will cease for his noble audience. Life in Elsinore will stand in arrest. And even the impatient Fortinbras, anxious to assume his royal functions, will be obliged to pause, like Hamlet, to hear of Hamlet. This pause in Elsinore to hear of Hamlet has already taken place in the Globe theater. For Shakespeare's somewhat less noble audience in London, Hamlet's staged story has also formed a juncture in the action of their lives. But within the breathing space of art, life itself begins to rouse and move to the compelling rhythms of the imagination. Gertrude says she has "no life to breathe" what Hamlet has said to her. To our infinite profit, however, Shakespeare has had life to breathe what the play *Hamlet* has said to us. In doing so he has both given life to and taken life from a breathing space.

Epilogue:

Presence and Absence

IN THE chapters above I have brought to bear on *Hamlet* various critical notions about presence and absence, negation, erasure, juncture, the synchronic and the diachronic, the concrete universal, and so forth. I have done so in eclectic fashion, not to enlist Shakespeare's play in the cause of a particular school of criticism or theory but simply as part of an effort to come to terms with the questions posed by the play. Still, the kind of criticism practiced here tends beyond explication and commentary toward the metadramatic, a search for the poetics in the play. Therefore it might be well at this point to append a brief unscientific postscript in an effort to bring some of the larger critical issues in *Hamlet* into closer focus.

I have sought to demonstrate that *Hamlet* is extraordinarily given to the *via negativa*. Here, perhaps more than anywhere else in his work, Shakespeare specializes in nonbeing and inaction, in forms of verbal suicide, in interruptive modes, in junctures and caesurae, in unnamings and unspeakings. To talk of these matters I began with Kenneth Burke and Bergson on the nature of the negative (Part II). A more fashionable starting point would have been the negative phrasings of structuralist and post-structuralist thought, with its heavy reliance on terms to which the pandemic prefix, *de*, has been attached. From Saussure's stress on the principle of differentiation in language, to Jacques Derrida's claim that the not-there asserts its presence, despite its absence, by leaving its "trace" on what is apparently there, the direction of this critical *via negativa* has been toward a celebration of absence within a linguistic system that has erased everything outside itself.[1] Abandoning the naive belief in a necessary connection between signifier and signified—as of course Plato had done in the *Cratylus*—entails abandoning also a metaphysics of "presence": the conception of an objective, unmediated, real world out there of which we can have direct knowledge. In the absence of such a world, and of any "transcendental Signified" above or behind it, what remains is what our symbolic systems project through the creative free play of fashioning fictions. By the same token, in literature, Derrida says, "the absence of an ultimate meaning opens an unbounded space for the play of signification."[2] As a result the critic is discouraged from the attempt to establish determinate meanings in a text or to assert the mimetic reality of its fictions. He should "decenter" such meanings and attend instead to what Julia Kristeva calls the "geno-text" of infinite semantic possibilities.[3]

Trying in an epilogue like this to do justice to these theories and to sort out their implications for *Hamlet* would cause a man to fall into a sadness, then into a fast, thence to a watch, thence into a weakness, thence to a lightness, and by this declension into an authentic melancholy adust. But perhaps I can suggest some of these implications by glancing again at a few of the issues treated earlier.

To begin with, it is clear that Shakespeare is by no means ingenuously committed to a poetics of presence in the Globe or to a metaphysics of presence in Denmark. Let us take Denmark first and especially the Ghost, which at first glance might almost have been created on the advice of Derrida. King Hamlet is funereally absent, King Claudius is complacently present, and the Ghost is somewhere in between, a "trace" of the dead king imposing its significance on a world in which it cannot fully reside. The Ghost emerges from silence and an unimaginable void to tell its secrets. However mystifying in itself, the Ghost assumes the role of demystifier. What it reveals quite subverts Hamlet's sense of reality and value; and although it is erased by the advent of dawn and by Hamlet's ritual swearings to silence of his companions, the Ghost's corrosive secrets live on, memorially inscribed within the "book and volume" of Hamlet's brain, where they "wipe away" all that "youth and observation copied there" (1.5.99 ff.).

As Hamlet's schoolboy conception of a world of received values and determinate meanings is exploded by the Ghost, so the audience's conception of *Hamlet* as a play grounded in mimetic conventions that yield an apparently unmediated experience of life in Denmark is exploded by Shakespeare. The playwright's conspicuous theatricalizing of Hamlet's world compels us to acknowledge the fact of dramatic mediation, to recognize that life in Denmark comes before us only as images in the refracting mirror of art. Indeed, Hamlet's story is as ghostly as the Ghost itself, in two senses. In the first place, somewhere in the dark backward and abysm of time, forever lost to us, lies the mythic experience of the Icelandic Amlothi. That experience passes through many voices before it reaches us in the theater—those of the skald Snaebjorn and Snorri Sturlason, of Saxo Grammaticus, of Belleforest, and, we assume, of Thomas Kyd. The original events of that oft-told story have secreted themselves somewhere on the far side of the fictions that depict them, rather as the distinctive nature of quarks hides from its representation in such metaphors as "flavors," "colors," "strangeness," and "charm" (both "naked" and "hidden"). But whereas there are nonmetaphorical ways of describing quarks, Shakespeare's play offers us for the theatrical moment our only access to the lost events it cannot help falsifying.

The loss of those events, if there were any, would be painful if we were much in love with history and Icelandic myth. But loss and disappearance are everywhere. For, in the second place, even the events we seem to have before us—the tragic doings of Denmark—are ghostly. The play is not what it is, or what it appears to be. It is not a presentation of the thing itself, the mimetic signified of Denmark, any more than Hamlet's schoolboy metaphysics of presence reflected the actual state of affairs at court. The theatergoer who submits himself to the ilusion of presence in the theater and weeps for Hamlet is pretty much at one with Polonius, who is moved by the actor who weeps for Hecuba. However, if Shakespeare deprives us of the Polonius-experience, in part by putting Polonius into the play, he compensates us for our loss by presenting us with the duplicitous but richly complex experience of Hamlet's world caught in the perspective of theatrical art. From this perspective *Hamlet* becomes a constitutive metaphor that simultaneously reflects and creates its own tenor. In this there is an overmastering presence.

Perhaps metaphor, with its dual commitment to likeness and unlikeness, is the best vehicle to convey the sense of presence and absence in *Hamlet*. Let me synopsize Hamlet's story again from this standpoint. I mentioned earlier that the transfer of the crown from "Hamlet" to "Hamlet" would have been metaphoric, based on the similarities of father and son. More than that, it would have involved a primitive, magical identity of the sort associated with mythic thought, so that the transfer of the crown would be not a sequence but a tautology. This kind of metaphor, the assertion of a consubstantial identity in the most naive form, is analogous to Hamlet's schoolboy metaphysics of presence, deconstructed by the Ghost, and to *Hamlet* as a drama of mimetic presence, deconstructed by Shakespeare.[4] However, this metaphoric sequence is intercepted by the actions of Claudius, whose accession to the throne is metonymic—based on the contiguity of brotherhood, which stresses not likeness but difference, as both Hamlet ("no more like my father/ Than I to Hercules") and the Ghost ("a wretch whose natural gifts were poor/ To those of mine") attest. In its assertion of metonymic difference, Claudius' kingship affirms the passage of time, the fact of death, and the sequentiality of the royal succession, all of which Hamlet's kingship, an assertion of metaphoric identity, would have denied.

So the dispossessed Hamlet loses a metaphoric role. But he quickly gains another. In place of the royal message he expected to receive as heir apparent, he receives from the Ghost an unforeseen message calling for revenge. As revenger he would also serve as a metaphoric substitute for his father, doing for him what that ghostly personage cannot do for himself. But although he at first welcomes this role, Hamlet subsequently proves no more apt for such revengeful business than the Ghost itself, in part because he has taken instruction from that deconstructive spirit. The Ghost has taught him how deviously discrete all signifiers and signifieds are in a Denmark bemused by Claudian lies and courtly seeming. As a result Hamlet defers the revengeful act that would identify him with his father and explores instead the field of differences that divides cosmetic appearance from ugly reality in Elsinore and separates himself from his dead father, his stepfather, his aunt-mother, the fair Ophelia, the world at large.

It appears, then, that Shakespeare has dramatized a movement beyond the ingenuous assumption of complete metaphoric identity to the recognition of absence or difference as the ground against which similarity is asserted. The world, after all, is far more unlike than it is like "an unweeded garden," death is more unlike than like an "undiscovered country," and Hamlet more unlike than like his father. During the middle of the play, while he defers the act that would confirm his likeness to his father, he explores their unlikeness, the dark side of the metaphor that links them.[5] For it is the negative dark side of that metaphor, with its implicit denial of identity, that seems to yield the Hamlet who is uniquely himself, disconnected not merely from his father but from everything "common." Thus in his role as madman and in the play of his soliloquies he repeatedly defines himself by what he is not. And at the same time that Hamlet says "Not this" of himself, Shakespeare's play says "Not this" of itself. The dramatic metaphor, which asserts that the play *is* its illusions, is deconstructed, leaving a gap of difference between Globe as vehicle and Denmark as tenor, between Burbage and Hamlet, between "is" and "seems." Shakespeare's metadramatic play upon these differences itself creates another difference, that which marks *Hamlet* off as wonderfully distinct from the *Ur-Hamlet* and other plays of the revenge tragedy genre.

But differentiation, which distinguishes and individualizes, also leaves us adrift in a space where nothing "is" and everything "is not." That is, we should acquire a very strange idea of both Hamlet and *Hamlet* if we were compelled to think of them exclusively in terms of what they are not. For what they "are" is as much dependent on what they are like as it is on what they are unlike. Shakespeare's tragic hero refuses to be identified wholly with his father, or with anyone else, and Shakespeare's play refuses to be identified wholly with the *Ur-Hamlet* or with other plays subsumed under the concept of revenge tragedy. But if we abstracted from Hamlet what he shares with his father and other men, and from *Hamlet* what it shares with other revenge tragedies, what would remain would be amorphous indeed. The virtue of metaphor, with its explicit assertion and implicit denial of identity, is that it acknowledges this fact.[6] This suggests that Hamlet the hero and *Hamlet* the play will not have completely defined themselves until they emerge from the blank space of metaphoric unlikeness and discover a more sophisticated illusion of metaphoric identity. Shakespeare, that is to say, is not content merely to deconstruct the world of his hero or the theatrical experience of his audience. He is not content merely with difference, negation, erasure, absence, with the "not to be" half of Hamlet's famous line. Hamlet and the play must both "be" and "not be." Thus, even before Hamlet emerges from the isolations of unlikeness and difference, his own negative declarations belie themselves. His advice to his mother, "Not this, by no means, that I bid you do," demonstrates that the "no" of conceptual negation cannot erase the poetic "yes" of imagistic verbal presence. So, too, with

Hamlet's wordplay. The pun, for instance, proves by nature too hospitable to evict the meanings Hamlet cannot abide; they all bed down together, for good and ill, in incestuous dalliance. Moreover, just as Hamlet particularizes himself in the process of generalizing, so he takes on a distinctive personal presence in the process of denying himself to be: he repeatedly says "I am not" in accents of eloquence that are uniquely Hamletic, in so individual a style that, as Shakespeare says of himself in Sonnet 76, "every word doth almost tell my name."

Thus Hamlet most specially "is" even while declaring himself "not to be." But his being acquires particular force in the graveyard, where the irreplaceable self encounters the commonness of death. Here he acknowledges, even proclaims, the union of his private and public identities: "This is I,/ Hamlet, the Dane!" This verbal affirmation of an identity that is both individual and generic, that acknowledges both difference and similarity, is confirmed in action when he kills the King both for himself and for his father. At that same point, Shakespeare, having established the individuality of his play, its metadramatic difference, acknowledges at last its generic nature as well by completing its revenge form.

Thus, out of the negative, the positive; out of not-being, being; and out of inaction, action. Similarly, out of deconstruction, reconstruction. For Hamlet's ability to be and to act is concurrent with his recovery of a sense of meaning in the world. His speeches about divinity and providence do not suggest that he has won his way to a new order of meaning, a distinctive philosophy. Nor, on the other hand, do they represent merely a bland return to the unexamined values of his schooldays, subsequently erased by the Ghost's story. The world and its values have not changed—only Hamlet. What he sees is as it was, and as it will be, but it has lost its dominion over him. He sees the world for what it is and for what it is not, and replies, "Let be." The cosmetic paintings, the lies, the corrosions of time on beauty and truth, the law's delay, the insolence of office—he knows them all; and perhaps most of all he knows the lingering dampness of the grave and the sound of the Gravemaker's shovel: "if it be not now, yet it will come." It is not, of course, to come, as we know; it is now.

The very stoic literalness of Hamlet's speech on readiness suggests a departure from his earlier soliloquies and wordplay. Language, it seems, is being stripped to its most elemental state. Beyond lie only the silent act and then silence itself. What has been so complexly verbal throughout the play, so fraught with ironies and erasures, negations, paradoxes, duplicities, all the evasive shifts and turns of speech, dissolves here at the end into the silence of the duel, into the pure presence of murderous physical action.[7] Shakespeare is returning us, as nearly as possible, to an unmediated presence, to naked action. In doing so he recapitulates Hamlet's own movement toward action in the play at large—that is, Hamlet's making his way toward the act of revenge through the nonactions of wordplay and play-acting. For the duel begins with "play"—the appearance of fighting mediated by the conventions of fencing and by the judge Osric, who tells us what has happened—and then it ceases to be play and becomes instead a mortal contest in which no conventions exist and no mediating judge is needed to tell who is hit and who is not. These actions speak directly, and as part of them Hamlet is intently, unambiguously, existentially engaged with his immediate world, even to the death.

To his own death, to the death of Laertes, and to the death of the King. The two killings of the King, as I suggested, do justice to the two-sidedness of metaphor, whose bifocal vision recognizes similarity within an ambience of difference. Of course naive metaphor, as in the equation of father and son in the Danish royal succession, does not assert similarity but identity. But Shakespeare has awakened us from that dream of primal unity in which things with the same names are the same. The identity now claimed is an illusion of identity that emerges from a full sense of difference as the principle by which Hamlet has distinguished himself from his father and by which *Hamlet* has distinguished itself from its dramatic forebear, the *Ur-Hamlet*, as well as other plays of like genre.

Unlike its earlier version, the final metaphor does not deny time and render the royal succession a tautology. By acknowledging difference it allows for metonymic contiguity, for sequence. Time might well have had a stop with the death of Hamlet, had Shakespeare elected to dramatize cessation at the expense of succession. But as it is, Hamlet's final act is to pass the kingship on to Fortinbras and so to keep open an avenue to the future. Like the Ghost, Fortinbras has hovered *in absentia* outside the ambit of the play, imposing his trace on Danish affairs from a distance. Yet at the end he too materializes, a figure of bluff physical immediacy and direct statement, to fill the absence left by Hamlet's "passage."

That the name of "Fortinbras," in which empty form the absent person of Fortinbras has existed until now, should take concrete shape as the brusque Norwegian prince is in keeping with the fact that Shakespeare's play has returned from the blank space of difference and absence to a subtler manifestation of identity and presence. That is, having deconstructed the theatrical reality of *Hamlet* for his audience, as the Ghost deconstructed the metaphysical reality of life in Denmark for Hamlet, Shakespeare now reinstates that reality but within a framework of illusion. As Hamlet is obliged to perform an act of revenge within a discredited world, so we as audience are obliged to perform an act of imagination within a theater of discredited illusions. As an act of poetic faith we must witness the tragic events of Denmark through the window of a theatrical performance that, as metaphor, cannot let us forget that it *is* a window and that what we see can be seen by no other means, for it is etched in the glass itself. Thus, on the one hand, the fact that Hamlet's killing of Laertes and Claudius begins as a "play" of swords, an illusion of fighting, causes the real fighting into which it degenerates to seem more immediate and convincing by contrast, as I suggested earlier. But if we pursue this transformation of illusion into reality one step further, we come upon an awkward reversal: the play-fighting by Hamlet and Laertes becomes real fighting between Hamlet and Laertes, which in turn becomes play-fighting by Burbage and (say) Heminges. Shift about as they will, neither the illusions of *Hamlet* nor its realities can escape their confinement within the master illusion of the play itself. And yet if we manage to see only Burbage and Heminges, only actors, in the Duel Scene, we shall have killed Hamlet and Laertes and Claudius in the theater quite as mercilessly as they are killed in court, a consummation hardly to be wished for. If Shakespeare has deconstructively split his play down the middle, dividing Burbage from Hamlet, Globe from castle, and the illusions of dramatic performance from the apparent realities of life in Denmark, he has subsequently reunited it. In the spirit of "To be and not to be," we are invited to register, not *either* Burbage *or* Hamlet, but the fusion of each in the other: Burbage-as-Hamlet, Globe-as-castle, illusion-as-reality. In this there are both presence and absence, for the play is a metaphor whose vehicle creates the tenor of which it is an image, and a sign whose signifier creates the signified which it means.

Notes

Prologue

1. The text that I am using is that of David Bevington, from *The Complete Works of Shakespeare*, 3d ed. (Glenview, Ill.: Scott, Foresman, 1980).

2. Sigurd Burckhardt, "Notes on the Theory of Intrinsic Interpretation," in *Shakespearean Meanings* (Princeton: Princeton University Press, 1968), pp. 285–313.

3. Geoffrey Hartman, *Beyond Formalism* (New Haven, Conn.: Yale University Press, 1970), p. 351. My juxtaposing Burckhardt and Hartman here is not meant to imply their theoretical agreement, although it was closer at the time Hartman wrote this book than it has become since.

4. I should mention here that Norman Rabkin some years back invoked the notion of "complementarity"—coined originally by Niels Bohr to account for the fact that certain phenomena in physics, especially light, can be conceptualized in mutually exclusive yet equally valid ways (light as undulatory, for instance, but also as corpuscular)—to designate the coexistence in Shakespearean drama of antinomic values (in *Hamlet*, for instance, the presence of reason and unreason as alternative defining features of both the hero and man himself). This is in *Shakespeare and the Common Understanding* (New York: The Free Press, 1966); but see also his later book, *Shakespeare and the Problem of Meaning* (Chicago and London: University of Chicago Press, 1981), which addresses itself to the apparent dissolution of meaning in Shakespeare occasioned by recent theories about deconstruction and undecidability. I should also call attention to an anticipation of my title by David Young, who says, in a line I fully endorse, "It is ironic, really, that 'To be or not to be' should be the most famous line in *Hamlet*, for its posing of opposites between which one is supposed to have to choose is far from characteristic of the play; 'to be *and* not to be' would be more to the point, given the tendency for opposite states to exist simultaneously within characters, situations, and the meanings of spoken lines" ("*Hamlet*, Son of *Hamlet*" in *Perspectives on Hamlet*, ed. William G. Holzberger and Peter B. Waldeck [Cranbury, N.J.: D.C. Heath, 1975], pp. 196).

5. Rebecca West, *The Court and the Castle* (New Haven: Yale University Press, 1958), reprinted in *Hamlet*, ed. Cyrus Hoy (New York: W.W. Norton, 1963), p. 265.

PART I. HAMLET: THE NAME OF ACTION

1. Ludwig Wittgenstein, *Tractatus Logico-Philosophicus*, trans. D.F. Pears and B.F. McGuiness (New York: Humanities Press, 1961), 3:26.

2. Gilbert Ryle, "The Theory of Meaning" in *Collected Essays: 1929–1968*, vol. 2 (New York: Barnes and Noble, 1971), p. 357.

3. Gilbert Ryle (see note 2 above), p. 358.

4. Michel de Montaigne, "Of Names," in *The Complete Works of Montaigne*, trans. Donald M. Frame (Stanford: Stanford University Press, 1958), p. 204.

5. Harry Levin, "Shakespeare's Nomenclature," in *Shakespeare and the Revolution of the Times* (New York: Oxford University Press, 1976), p. 68.

6. Abraham Fraunce, *The Lawiers Logike* (London, 1588); cited by Sister Mirriam Joseph, *Shakespeare's Use of the Arts of Language* (New York: Columbia University Press, 1947), p. 312.

7. On the Maori, see Elsdon Best, *The Maori* (Wellington, N.Z., 1924), vol. 2, p. 452. The status of Homeric man is dealt with by Bruno Snell in *The Discovery of the Mind*, trans. Thomas Rosenmeyer (Cambridge, Mass.: Harvard University Press, 1953). Snell ascribes the discovery of the self to the Greek lyric poets (see his third chapter). In his *Preface to Plato* (Cambridge, Mass.: Harvard University Press, 1963), Eric A. Havelock says "The doctrine of the autonomous psyche is the counterpart of the rejection of the oral culture [that had been dominant in Greece prior to the fifth century]" (p. 200).

8. This identification of Hamlet with his father, as suggested by his being both namesake and deputy, is given a psychological turn by Theodore Lidz in his *Hamlet's Enemy* (London: Vision Press, 1976), p. 48:

> Mourning is a complicated psychological process. One aspect concerns the mourner's heightened identification with the deceased; in a sense, the mourner unconsciously attempts to keep the deceased alive by becoming like him, by incorporating the deceased into his own self. Insofar as the mourner fails to live up to the idealized model of the mourned person, he punishes himself by self-derogation.

9. For a further discussion of the dynastic parallels, see chapter 34.

10. Terence Eagleton, in an excellent discussion of characters as autonomous subjects and maneuverable objects, takes a similar view of Laertes in *Shakespeare and Society* (New York: Schocken Books, 1967), p. 61.

11. T.S. Eliot, "Hamlet and his Problems," in *Selected Essays: 1917–1932* (New York: Harcourt, Brace, 1932), p. 124.

12. See Ernest Jones, *Hamlet and Oedipus* (New York: W.W. Norton, 1949). Other full-length psychoanalytic studies of fairly recent vintage are K.R. Eissler's exhaustive *Discourse on Hamlet and "Hamlet": A Psychoanalytic Inquiry* (New York: International Universities Press, 1971); Theodore Lidz's *Hamlet's Enemy: Madness and Myth in Hamlet* (New York: Vision Press, 1975); and Avi Erlich's *Hamlet's Absent Father* (Princeton: Princeton

University Press, 1977). Norman Holland has a perceptive discussion of the relevance of psychoanalysis to literature, especially to Shakespeare, in *Psychoanalysis and Shakespeare* (New York: McGraw-Hill, 1964); and Paul Gottschalk supplies a review of "Psychological Interpretation: Main-Line Criticism Since Bradley" in his helpful *The Meanings of Hamlet: Modes of Literary Interpretation Since Bradley* (Albuquerque, N.M., University of New Mexico Press, 1972), pp. 78–101.

13. See Kenneth Burke, *A Grammar of Motives & A Rhetoric of Motives* (Cleveland and New York: The World Publishing Company, 1962), p. 249.

14. Of course all language is by nature abstract and hence unable to deposit concrete referents before us as Swift's Grand Academicians of Lagado could do. Even the particularizing of "the quarterback for the Rams wearing number twelve" is accomplished by a series of abstract terms. Nevertheless, within a context of indispensable abstraction, we can move down a scale from abstractness toward concrete particularity in dealing with things (e.g., abstract nouns, common nouns, proper names) in a way we cannot do in dealing with actions.

15. In a brief essay called "Who Thinks Abstractly?" Hegel makes a distinction between judging people by what they are as opposed to what they do. He says, "This is abstract thinking: to see nothing in the murderer except the abstract fact that he is a murderer, and to annul all other human essence in him with this simple quality" (*Hegel: Texts and Commentary*, trans. Walter Kaufmann [New York: Anchor Books, 1966], pp. 116–17). Harold Rosenberg makes a similar distinction between "personality," which represents the felt unity of evolving organisms, and "identity," or "the character defined by the coherence of his acts" ("Character Change and the Drama" in his *The Tradition of the New* [New York: Horizon Press, 1965], pp. 135–53). See also regarding the defining properties of action and its consequences Hannah Arendt's section on "Action" in *The Human Condition* (Chicago: University of Chicago Press, 1958), pp. 175–243.

16. Maynard Mack, "The World of *Hamlet*," *The Yale Review* (1952) 41:502–23. Like so many others, I am indebted to Professor Mack's brilliant study of the play.

17. To make an important point even clearer, I would depart from the usual textual procedure here, which places a comma after "Taint not thy mind" and continues "nor let thy soul contrive/ Against thy mother aught," in favor of the fuller stop of the semicolon that appears after "mind" in the Folio text. The semicolon makes a more sweeping demand for mental purity on Hamlet's part than the comma does, and is thus more in keeping, it seems to me, with the stress in the play upon a contagion that includes but is not restricted to Gertrude's infidelities. Construing the Ghost's "Taint not thy mind" as a general injunction does not mean that we should disregard the association of Hamlet with Orestes, for instance, or with Nero (3.2.393)—an association that is well argued by Harold Fisch in *Hamlet and the Word* (New York: Frederick Ungar, 1971), pp. 78–100—but merely that we should consider the matricidal theme as part of a larger context of contamination.

18. T.S. Eliot (see note 11 above), p. 125.

19. Anonymous, *Some Remarks on the Tragedy of Hamlet, Prince of Denmark, Written by William Shakespeare* (London, 1736), pp. 32–33.

20. In "Hamlet When New" (*Sewanee Review* [1953], 41:15–42) William Empson argues that Shakespeare has Hamlet call attention to his delay in order to provide a lightning rod for the audience's impulse to laugh at the creaky revenge tragedy structure (which

Empson supposes to have gone out of fashion some ten years earlier) and to make theatricality itself a theme of the play.

21. Geoffrey Hartman has a fine discussion of various examples of "overspecified ends and indeterminate middles" that force us to the act of interpretation, in "The Voice of the Shuttle: Language from the Point of View of Literature" in *Beyond Formalism* (New Haven, Conn.: Yale University Press, 1970), pp. 337–55.

22. Thomas Wilson, *The Arte of Rhetorique* (1553), ed. G.H. Mair (Oxford: Oxford University Press, 1909), p. 181.

23. Thomas Wilson (see note 22 above), p. 182.

24. Samuel Johnson, from "Notes on *Hamlet*" in *Jonson on Shakespeare*, ed. Walter Raleigh (London: Oxford University Press, 1908), p. 196.

25. William K. Wimsatt, Jr. and Cleanth Brooks, *Literary Criticism: A Short History* (New York: Alfred A. Knopf, 1964), p. 627. Ransom's theory appears in various of his articles and books but most clearly perhaps in "Criticism as Pure Speculation" in *The Intent of the Critic*, ed. Donald A. Stauffer (Princeton: Princeton University Press, 1941) and in "The Concrete Universal," *The Kenyon Review* (Summer 1955), later reprinted in Ransom's *Poems and Essays* (New York: Vintage Books, 1955).

26. These metatheatrical matters, with a stress on plays within plays, are taken up by Robert J. Nelson, *Play within a Play* (New Haven: Yale University Press, 1958); Anne Righter, *Shakespeare and the Idea of the Play* (London: Barnes and Noble, 1962); Lionel Abel, *Metatheater* (New York, 1963); Thomas Stroup, *Microcosmos, the Shape of the Elizabethan Play* (Lexington, Ky.: University of Kentucky Press, 1963); Robert Egan, *Drama Within Drama* (New York and London: Columbia University Press, 1972, 1974); and Jackson I. Cope, *The Theater and the Dream* (Baltimore: Johns Hopkins University Press, 1973); and Sidney Homan, *When the Theater Turns to Itself* (East Brunswick, N.J.: Bucknell University Press, 1981). Shorter studies include Maynard Mack, "The World of *Hamlet*," *The Yale Review* (1952), vol. 41; C.G. Thayer, "*Hamlet:* Drama as Discovery and as Metaphor," *Studia Neophilologica* (1956), vol. 28; Richard Foster, "*Hamlet* and the Word," *UTQ* (1961), vol. 30; Charles R. Forker, "Shakespeare's Theatrical Symbolism and Its Function in *Hamlet*," *SQ* (1963), vol. 14; Maurice Charney's section on "Art, Acting, and the Theater" in his *Style in Hamlet* (Princeton: Princeton University Press, 1969), pp. 137–53; Harold Fisch's chapter "All the World's a Stage" in his *Hamlet and the Word* (New York: Frederick Ungar, 1971), pp. 119–28; Nigel Alexander's chapter "Poison in Jest: the Play Scene" in his *Poison, Play, and Duel* (Lincoln, Neb.: University of Nebraska Press, 1971), pp. 91–118; and Lawrence Danson's chapter "Hamlet" in *Tragic Alphabet* (New Haven and London: Yale University Press, 1974), pp. 22–49. A perceptive awareness of such issues also underlies the analyses of the play by Bridget Gellert Lyons in her *Voices of Melancholy* (London: Routledge and Kegan Paul, 1971), pp. 77–112; by Rosalie Colie in *Shakespeare's Living Art* (Princeton: Princeton University Press, 1974), pp. 208–42; and by Kirby Farrell in his *Shakespeare's Creation* (Amherst: University of Massachusetts Press, 1975), pp. 163–91.

27. Keats's famous definition of Negative Capability appears in his letter to George and Thomas Keats on Sunday, December 21, 1817; see Lionel Trilling, *The Selected Letters of John Keats* (New York: Farrar, Strauss, and Young, 1951), p. 92.

28. That Hamlet's "antic disposition" has been discarded is indicated not only by

his new apparel but by his new language, as Bridget Gellert Lyons so well notes in *Voices of Melancholy* (London: Routledge and Kegan Paul, 1971), p. 105:

> On entering the graveyard [Hamlet's] first remark (with its obvious pun on "grave") is tantamount to a denial of his behaviour in earlier episodes: "has this fellow no feeling of his business, that 'a sings in grave-making?" (V,i,64). It is now the Clowns who do most of the fooling, and the quibbles they indulge in, based on their inability or unwillingness to understand what their interlocutor is saying, are not in fact very different from Hamlet's more deliberate quibbling earlier. Yorick's skull, the focus of the early part of the scene, is the most visible token of Hamlet's abandonment of the comic mode. His meditations about death and the passing of all things include thoughts about the transitoriness of the jokes and antics of the Court Jester. To the extent that Hamlet himself has been fulfilling the role of jester, the scene is a comment, visual and verbal, on the whole "antic disposition."

As Professor Lyons suggests, when Hamlet unburies the skull of the jester Yorick, he buries the jester in himself. It is significant that the court of Claudius will permit no official jester. When the lie wears ermine, the truth cannot go even in motley.

29. Maynard Mack (see note 16 above), p. 520.

30. This notion of a shift is advanced by Harold Rosenberg in a thoughtful discussion of character change in the drama (see note 15 above).

31. The opening lines of Act 5, Scene 2 make a gesture toward accounting for this by having Hamlet imply that he has already, perhaps before the Graveyard Scene, told Horatio something about his adventure ("You do remember the circumstance?"). This could be, however, merely a reference to Hamlet's letter to Horatio in Act 4, Scene 6, telling about his escape at sea with the pirates. But in either case it is clear from Horatio's astonished responses to Hamlet's story that, whatever he has heard or read before, he has not encountered *these* details. And that leaves us where we were—wondering why.

32. Like the divine playwright, Shakespeare is not so preoccupied with the angels up above that he forgets the sparrows down below. His dramatic universals, like Aristotle's and unlike Plato's, are fully entangled in the mortal stuff of the world. If at the end there are glimpses of theological transcendence in Hamlet's references to divinity and providence, nevertheless it is a transcendence that is inextricably involved in acts of murder (Hamlet's "salvation" at sea is inseparable from his Claudius-like dispatching of Rosencrantz and Guildenstern), just as his fulfillment of his revengeful assignment is unavoidably contaminated by his poisonous means. In chapters 10 and 11 below I suggest that the redundancy of Hamlet's revenge on Claudius represents Shakespeare's maneuvering between the concrete and the universal in Hamlet and in *Hamlet*.

33. Sir Walter Raleigh, *History of the World* (1681 ed.), p. xxi.

34. During the presentation of "The Murder of Gonzago" Hamlet dominates the court by virtue of his special knowledge of the play, particularly as rewritten by himself. For the first time he has the King at his mercy. Immediately afterwards in the Prayer Scene he again has the King at his mercy—both his life and, as he feels, his afterlife as well. This idea of Hamlet contesting with Claudius for control over life in Denmark is ably examined by Thomas F. Van Laan in his *The Idiom of Drama* (Ithaca and London: Cornell University Press, 1969), pp. 241–42, 248–50.

35. This is not to suggest that Hamlet remains untainted by the Claudian world or his own actions. He has, after all, abandoned and insulted Ophelia, murdered her father (hence contributing to her madness and death), and sent his school-fellows to sudden death "Not shriving time allowed" (5.2.47). But Hamlet is in the Antigone situation: damned if he does, damned if he doesn't. Or with Thomas Becket in Eliot's version he might well ask, "Can I neither act nor suffer/ Without perdition?" In fact, that is pretty much what he does ask in the "To be or not to be" speech: Is it nobler to suffer this world or to take arms against its evils? Stoic suffering or heroic action: don't they both leave one in the perdition of living on?

36. See note 28 above for Bridget Gellert Lyons' remarks about Hamlet's doffing his "antic disposition." Moreover, the present tense verbs from line 232 on in this speech are subjunctive hypotheses, not statements of present fact.

37. Hamlet is dealt with rather sharply by Lee Sheridan Cox in *Figurative Design in Hamlet* (Columbus, Ohio: Ohio State University Press, 1973) for his "deliberate and *superfluous* hypocrisy here in portraying himself as victim to the man whose father he has slain" (p. 138). By bringing Hamlet's own earlier sentiments to testify against him, Cox argues forcefully against the usual view that Hamlet makes a gracious gesture of apology here. It is not, perhaps, one of Hamlet's better moments, but I wonder if we shouldn't lend more weight to Laertes' own response, which is to be "satisfied in nature" and to receive "offered love like love" while remaining aloof on the question of honor until "elder masters" have adjudicated. And of course our attitude toward Hamlet's lack of diplomacy is tempered by our awareness of the even greater hypocrisy of Laertes, standing on points of honor while he scrutinizes the swords in search of the unbated one. Interestingly, Mr. Cox arrives at the view that Hamlet the son and revenger is at least metaphorically "mad"(p. 140) in a way somewhat like that argued here.

38. Moreover, there is a certain justice in Claudius' first death being inflicted by Hamlet the individual, since Hamlet owes his individuality in some measure to Claudius. That is, had Hamlet become King of Denmark in the normal course of affairs, after the death of his natural father, we should have had King Hamlet succeeded by King Hamlet—not so much a "succession" as merely a substitution that would symbolically perpetuate the reign of the first Hamlet. On this view Hamlet would have "become" his royal father without having an opportunity to discover his own identity. But when Claudius intervenes in the royal succession Hamlet retreats in grief and "madness" into the sanctuary of the private self. It is fitting, then, that Claudius is first attacked, not by the "son" whom he has feared and sought to murder, but by the "self" he helped create but never knew existed.

39. In a fine article entitled "Hamlet, Duellist" (*UTQ* [October 1969], 34[1]) Sheldon Zitner notes that Hamlet's "innovation as a revenge-play lies precisely in its exploration of the self-consciousness of the avenger" (p. 5). Some of his concluding remarks, incidentally, merge eloquently with the interpretation of the play presented in this chapter:

> Under Shakespeare's hand, the crisis of aristocratic power and the code which expressed it became hieroglyphs of a universal crisis and its forms: the crisis of young manhood and the course of maturation. Hamlet is no more his father's child, nor yet his own man—he is at home among neither kin nor kind; his are the deracinations of all young men. . . . All young men have a father in some sense lost, a mother (dare one say it without arousing sensitivities to Freud) to whom they cannot return in innocence. And

all are plagued by history, by contradictory injunctions from the noble dead, impossible to realize or to ignore. The past pops in between their election and their hopes, offering only obligations where they would grasp the heroic freedom that as children they grew to believe was their patrimony. (pp. 16–17)

PART II. TO BE AND NOT TO BE: THE RANGE OF NEGATION IN HAMLET

1. Maynard Mack, "The World of *Hamlet*," *The Yale Review* (1952), 41:504.

2. Kenneth Burke, *Language as Symbolic Action* (Berkeley and Los Angeles: University of California Press, 1968), p. 419. Bergson and Burke were partly anticipated by Plato in his response to the Sophists. The Sophists argued that if we speak of that which is not we speak of nothing—which is true enough when we say, for instance, "The furry lizard does not exist"—and then claimed that they themselves could not be convicted of falsehood or error because these involve saying the thing that is not, and what is not is nothing. Plato, on the other hand, held that Not-Being has Being, on the grounds that if we say that A is not B we are not saying that A is nothing but merely that it is something other than B. The difference depends on whether we take the verb "is" as an existence-claim or merely as a functional copula. See Plato's *Sophist*, especially 257b–258e, in *The Collected Dialogues of Plato*, ed. Edith Hamilton and Huntington Cairns (Princeton: Princeton University Press, 1961).

3. Kenneth Burke (see note 2 above), p. 430. Joseph Vendryès makes the same point a few years after Bergson: "To make the reader feel the contrary of a given impression, it is not enough to bracket the words which convey it with a negative. For that is not the way to suppress the impression one wishes to avoid: one evokes the image, while thinking he is banishing it." That is, as Maurice Merleau-Ponty sums up, "There are denials that affirm." See Vendryès, *Le Langage* (Paris, 1923), pp. 159–60; cited by Merleau-Ponty in *The Prose of the World*, ed. Claude Lefort, trans. John O'Neill (Evanston, Ill.: Northwestern University Press, 1973), p. 30. In stressing the positive aspects of negation, these writers invert the position of Ferdinand de Saussure, who stresses the negative, differential aspect of apparent positive values in language: "there are only differences, with no positive terms" (*Course in General Linguistics*, trans. Wade Baskin [New York: McGraw-Hill, 1966], p. 120).

4. Kenneth Burke (see note 2), p. 420.

5. Plato, *Cratylus*; Saussure, *Course in General Linguistics*.

6. *Saint Thomas Aquinas: Philosophical Texts*, trans. Thomas Gilby (New York: Oxford University Press, 1960), p. 168, from III *Contra Gentes*, 7. See also Sigurd Burckhardt, *Shakespearean Meanings* (Princeton: Princeton University Press, 1968), p. 271.

7. Romans 7:7. Also Montaigne quotes Ovid, "She who does not, because forbidden, really does" (*The Complete Works of Montaigne*, trans. Donald M. Frame, [Stanford: Stanford University Press, 1948], p. 478); and I recall having read, though I cannot find where, that at an early point in Greek or Roman history the incidence of patricide rose sharply after the promulgation of a law against it. More authoritative yet, somewhere in the works of Groucho Marx is the statement, "I was a teetotaler until Prohibition."

8. Like picturable things, actions have no negatives. Just as we do not have words

like "un-dog" or "non-book" (though we have "nonentity" for an unpicturable abstraction), so we do not have words like "un-reading" or "non-walking," presumably because if we are *not* engaged in these actions we *are* engaged in others. As Plato argued, if the dog is not black, it is white or brown or whatever; it is not nothing. The negative form of an action is what we take to be its opposite—"going" vs. "coming," "moving" vs. "standing," "working" vs. "playing," etc.—all of which are positive terms.

9. See René Girard, *Violence and the Sacred*, trans. Patrick Gregory (Baltimore and London: Johns Hopkins University Press, 1977); originally *La Violence et le sacre* (Paris, 1972); pp. 49–51.

10. *The Sources of Hamlet*, Sir Israel Gollancz (London: Oxford University Press, 1926), reprinted by Octagon Books (New York, 1967), p. 103. Or *Narrative and Dramatic Sources of Shakespeare*, vol. 7, Major Tragedies, ed. Geoffrey Bullough (London and New York: Oxford University Press, 1973), pp. 63–64.

11. Critics have been quick to cite the serpentine allegorical truth of Claudius' lie, and in some cases to regard the entire play as a reenactment of the Fall and subsequent Redemption by a Christ-like Hamlet. For the Christian framework of the play, see Roy Walker, *The Time Is Out of Joint* (London: Andrew Dakers, 1948); G.R. Elliott, *Scourge and Minister* (Durham, N.C.: University of North Carolina Press, 1951); Bertram Joseph, *Conscience and the King* (London: Chatto and Windus, 1953), pp. 130–51; V.K. Whitaker, *Shakespeare's Use of Learning* (San Marino, Calif.: The Huntington Library, 1953), pp. 262–74; Paul N. Siegel, *Shakespearean Tragedy and the Elizabethan Compromise* (New York: New York University Press, 1957), pp. 99–118; H.S. Wilson, *On the Design of Shakespearian Tragedy* (Toronto: University of Toronto Press, 1957), pp. 31–51; Irving Ribner, *Patterns in Shakespearian Tragedy* (London: Methuen, 1960), pp. 65–90; William B. Toole, *Shakespeare's Problem Plays* (The Hague: Mouton, 1966), pp. 98–121; and Harold Fisch, *Hamlet and the Word* (New York: Frederick Ungar, 1971), esp. pp. 101–18.

12. Compare, for instance, Iago's remark as he is preparing his lies for Othello: "I'll pour this pestilence into his ear" (2.3.362). The equation of Claudius' ear-poisoning with lying and verbal contamination generally was first made, I believe, by Sigurd Burckhardt in "The King's Language: Shakespeare's Drama as Social Discovery," *Antioch Review* (1961), 21:369–87. Several critics have followed Wolfgang Clemen in recognizing that the disease imagery noted by Caroline Spurgeon in *Shakespeare's Imagery* (Cambridge: Cambridge University Press, 1935) results from the spreading poison first placed in Old Hamlet's ears. See, for instance, Clemen's *The Development of Shakespeare's Imagery* (Boston: Hill and Wang, 1951), pp. 113 ff.; Maynard Mack, "The World of *Hamlet*," *The Yale Review* (1952), vol. 41; M.M. Mahood, *Shakespeare's Wordplay* (London: Methuen, 1957), p. 118; and Maurice Charney, *Style in Hamlet* (Princeton: Princeton University Press, 1969), pp. 34 ff. In the line of Burckhardt, with a stress on verbal poison, see Terence Hawkes, *Shakespeare's Talking Animals* (London: Edward Arnold, 1973), pp. 105–26; T. McAlindon, *Shakespeare and Decorum* (London: Barnes and Noble, 1973), pp. 48–52; and Lee Sheridan Cox, *Figurative Design in Hamlet* (Columbus, Ohio: Ohio State University Press, 1973), pp. 35–52. Nigel Alexander's comprehensive *Poison, Play, and Duel: A Study in Hamlet* (Lincoln, Neb.: University of Nebraska Press, 1971) touches all of these bases.

13. See Terence Hawkes (see note 12), p. 113. Earlier on in his excellent chapter on *Hamlet*, Hawkes supplies a detailed analysis of each of the phases of this descent of

speech. See also Lawrence Danson, *Tragic Alphabet* (New Haven: Yale University Press, 1974), pp. 22–49, for a well conducted discussion of verbal deterioration in Elsinore and of the close relation between verbal and actional expression.

14. In *Shakespeare and Decorum* (see note 12), T. McAlindon makes this very interesting point:

> Incest being the most fundamental disorder in human relationships, it is appropriate that the incestuous nature of his mother's second marriage should provoke Hamlet's most memorable riddles. The play thus provides striking literary evidence in favour of the anthropological theory that there is a correlation in mythic thought between incest and riddling. (p. 59)

For the anthropological theory mentioned, McAlindon cites Claude Levi-Strauss, *The Scope of Anthropology*, trans. S.O. and R.A. Paul (London: Cape, 1967), pp. 34–39. *Hamlet*, in which the remarriage produces only dual identities—aunt-mother, uncle-father, and nephew-son—might not be as fertile a source of riddles as, say, *Oedipus Tyrannos*, which creates in Oedipus the triple identity of son-husband-father.

15. Hamlet has occasionally been taken to task for this romantic withdrawal from the "complexities of adult living," as L.C. Knights puts it in "Prince Hamlet," *Explorations: Essays in Criticism* (London: Chatto and Windus, 1946), reprinted in *Discussions of Hamlet*, ed. J.C. Levenson (Boston: D.C. Heath, 1960), p. 80. Such criticism seems better suited to Romeo than to Hamlet, who does, after all, abandon the sanctuary of the private self and engage with the corruptive complexities of Elsinore.

16. I should quote Sister Mirriam Joseph at this point: "The word pun, according to the *Oxford English Dictionary*, appeared first about 1660. What we call puns rhetoricians of the Renaissance subdivided into a number of figures which from ancient times were regarded as adornments." See *Rhetoric in Shakespeare's Time* (New York: Columbia University Press, 1947), p. 340.

17. M.M. Mahood, *Shakespeare's Wordplay* (London, 1957), p. 112.

18. Wolfgang Clemen (see note 12), p. 110. Coleridge speculated that Hamlet's earliest puns were "either due to 1. exuberant activity of mind, as in Shakespeare's higher comedy; or 2. imitation of it as a fashion . . . or 3. contemptuous exultation in minds vulgarized and overset by their success, like Milton's Devils; or 4. as the language of resentment . . . or lastly, as the language of suppressed passion, especially of hardly smothered dislike." See *Coleridge's Writings on Shakespeare*, ed. Terence Hawkes (New York: G. P. Putnam's Sons, 1959), p. 145. T.S. Eliot, apparently keying on Coleridge's "language of suppressed passion," says "The levity of Hamlet, his repetition of phrase, his puns, are not part of a deliberate plan of dissimulation, but a form of emotional relief. In the character Hamlet it is the buffoonery of an emotion which can find no outlet in action; in the dramatist it is the buffoonery of an emotion which he cannot express in art." See "Hamlet" in Eliot's *Selected Essays 1917–1932* (New York: Harcourt, Brace, 1932), pp. 125–26.

19. See Maurice Charney re Claudius' "brayings," *Style in Hamlet* (Princeton: Princeton University Press, 1969), pp. 25–26, and T. McAlindon re "noise" (see note 12), pp. 48–52.

20. T. McAlindon (see note 12), p. 50.

21. Two questions about the Ghost: What actually is its identity? and Why does

Hamlet doubt that it is his father's spirit? Concerning the first question there has been much throwing about of scholarly brains. The Ghost has been called an hallucination by W.W. Greg ("Hamlet's Hallucination," *MLR* [1917], vol. 12), an angel by Sister Mirriam Joseph ("*Hamlet*, A Christian Tragedy," *SP* [1962], vol. 59), a devil by R.W. Battenhouse and Eleanor Prosser ("The Ghost in *Hamlet*: a Catholic 'Linchpin'?" *SP* [1951], vol. 48, and *Hamlet and Revenge* [Stanford, 1967]), and "a compound of the Senecan revenge ghost, the Catholic purgatorial spirit, and the popular graveyard spook" by P.N. Siegel ("Discerning the Ghost in *Hamlet*," *PMLA* [1963], vol. 78). The entire matter of the Ghost and its critics is well summarized by Paul Gottschalk in his *The Meanings of Hamlet*, Appendix A (Albuquerque, N.M.: University of New Mexico Press, 1972), pp. 142–51. Very few people in the theater, I suspect, question the Ghost's claims to be the spirit of Hamlet's father, any more than Hamlet himself does during and shortly after their meeting. If we regard the Ghost as something else—angel, devil, whatever—we forfeit the ironic contrasts between this meeting and the relation of Laertes to *his* father in the deliberately juxtaposed scenes before and after this. As Susan Snyder perceptively remarks,

> We see one world in which Hamlet is trying to evade a stepfather he despises and to bear a terrible burden of revenge imposed on him by his own father from beyond the grave. Juxtaposed is a smaller, safer world in which Laertes evades *his* omnipresent father by retreating to Paris and Polonius tries to reach him, not across the abyss of eternity but over the measurable distance between Denmark and France, not with dread commands but with "assays of bias" and "slight sullies" (*The Comic Matrix of Shakespeare's Tragedies* [Princeton: Princeton University Press, 1979], p. 110).

As for the second question, Hamlet's immediate response to the Ghost in 1.5 is an unquestioning conviction that it is his father's spirit; only after a lapse of some time does he have doubts and wonder about devilish doings. Why? His imaginative transformation of the Ghost's identity from "father's spirit" to "possible devil" could be construed as a revelation of his suppressed Oedipal fear of his father. Perhaps more to the point, however: his own identity is as much in question as that of the Ghost. If the Ghost *is* his father's spirit, then Hamlet's identity must become that of dutiful revengeful son. But if the Ghost is not his father's spirit, then Hamlet's identity is that of independent inquiring adult. By doubting the Ghost, Hamlet provides himself with an excuse for deferring the act of revenge and for deferring his commitment to a particular identity, either son or self. It is typical of Hamlet that he prefers doubt to commitment, and the indeterminate zone of "maybe" to the fixed categories of "yes" or "no."

22. See chapter 22 below, "Beyond Negation." That providence takes care of those who do not presume too vigorously to take care of themselves is suggested also by the case of Fortinbras, who achieves his "revenge"—his recovery of the lands forfeited by his father to Old Hamlet—not by a frontal assault on Denmark but by submitting to his uncle's will, forgoing revenge, and acting for himself in Poland. When he returns to Denmark, the lands he had first thought to steal are presented to him almost as a gift, somewhat as Hamlet's opportunity to kill the King is presented to him. On the role of providence in the play, see Robert G. Hunter, *Shakespeare and the Mystery of God's Judgments* (Athens, Ga.: University of Georgia Press, 1976), pp. 101–26.

23. John Dover Wilson, *What Happens in Hamlet* (Cambridge: Cambridge University Press, 1935).

24. Harold Jenkins, "Hamlet and Ophelia," *British Academy Proceedings* (1963), 49:135–51.

25. Oedipal theorists sometimes cite this fact in evidence of Hamlet's unconscious desire to kill his father. For instance, K.R. Eissler—in *Discourse on Hamlet and* Hamlet: *A Psychoanalytic Inquiry* (New York: International Universities, 1971), pp. 134–37—rejects the notion that by substituting "nephew" for "brother" in "The Mousetrap" Hamlet is forecasting his eventual murder of Claudius, on the curious grounds that such a thought is acceptable to Hamlet. That is, Eissler regards the Inner Play from a psychoanalytic standpoint as a "dream-within-a-dream" and assumes that Hamlet's contributions to this dream should, as in ordinary dreams, express unacceptable desires in an acceptable fashion—through symbolism, condensation, projection, etc. But since the idea of killing Claudius is perfectly acceptable to Hamlet during waking life, he would feel no need to express it in the dream-within. What Hamlet really seeks to express, Eissler argues, is the consciously unacceptable desire to kill his father, and he renders this desire acceptable within the "dream" by labeling the dream-killer a nephew instead of a son.

I suspect that William of Occam would advance upon this argument with his razor at the ready, pointing out that the category of "dream-within-a-dream" involves a needless multiplication of entities. The Inner Play, that is, is just what it calls itself, a play, and specifically one that has been altered to catch the conscience of the King. As such, it is not obliged to express what is unacceptable in Hamlet's mind but what is unacceptable in the King's mind. Thus Hamlet, in identifying the play-murderer, substitutes "nephew," not for "son," but for "brother." If that lets Claudius off the hook a bit as far as the past murder is concerned, it places him squarely on it again as far as the future revenge for that murder is concerned.

26. Especially, as Maurice Charney points out, the verbal style of the royal statist Claudius, in which "*as*-es" multiply almost as rapidly as they do in Hamlet's parody—eight times, for instance, in 4.7. See Charney (note 19), p. 281*n*.

27. Harry Levin says of Hamlet's use of his father's signet, "It was the act of a king" (*The Question of Hamlet* [New York: Oxford University Press, 1959], p. 94); and in her fine chapter on the play in *Shakespeare's Living Art* Rosalie Colie says, "Hamlet waits in the wings of a kingship rightfully his, and polishes off his school-fellows by playing a king, with his father's seal signing their death-warrant" (Princeton: Princeton University Press, 1974, p. 222).

28. Susan Snyder has some particularly acute observations about the threat of the Gravemaker "to the irreplaceable 'I' of tragedy" and about the entire scene as constituting a "comic-absurd challenge to heroic individuality" to which Hamlet replies with "a series of self-affirmations." See her excellent study, *The Comic Matrix of Shakespeare's Tragedies* (Princeton: Princeton University Press, 1979), esp. pp. 125–30. In *The Music of the Close: The Final Scenes of Shakespeare's Tragedies* (Lexington, Ky.: University of Kentucky Press, 1978), Walter C. Foreman, Jr. presents a perceptive discussion of the Graveyard Scene; see pp. 83–98. See also Yasuhiro Ogawa's excellent article, "Grinning Death's-head: *Hamlet* and the Vision of the Grotesque" (*Northern Review* [1980], no. 8), for a discussion of the play as a "*Todesschmerz*."

29. The moral and theatrical subtleties involved in bringing Hamlet to his act of revenge are ably discussed by Richard T. Brucher in "Fantasies of Violence: *Hamlet* and *The Revenger's Tragedy*," *SEL* (Spring 1981), 21(2):257–70. See also Fredson Bowers,

"Hamlet as Minister and Scourge," *PMLA* (1955), 70:740–49; Maurice Charney, "The Persuasiveness of Violence in Elizabethan Plays," *RenD* (1969), n.s. 2; Harold Skulsky, "Revenge, Honor, and Conscience in *Hamlet*," *PMLA* (1970), 85:78–87; Eleanor Prosser, *Hamlet and Revenge*, 2d ed. (Stanford: Stanford University Press, 1971); R.A. Foakes, "The Art of Cruelty: Hamlet and Vindice," *ShS* (1973), vol. 26; and Michael Cameron Andrews, "*Hamlet*: Revenge and the Critical Mirror," *ELR* (Winter 1978), 8(1):9–23.

30. Kenneth Burke, *A Grammar of Motives and A Rhetoric of Motives* (Cleveland and New York: World, 1962), pp. 1–15.

31. In Shakespeare's imagination, fullness and vividness of individual being seem to be associated with characters in a pre-actional state—with characters who are contemplating but not yet committed to action, like Brutus and Macbeth, or perhaps with characters whose style of life constitutes a deferment of responsible action, like Antony in Egypt, Hal in Eastcheap, Falstaff, Sir Toby Belch—presumably because the compulsion to act inhibits the freedom to be. In her excellent study, *Dream in Shakespeare* (New Haven and London: Yale University Press, 1974), Marjorie B. Garber associates this pre-actional (or even pre-actual) state in Hamlet with dream, a retreat from outer reality into a subjective world of the imagination (which has its outer counterpart in what I have called madplay and wordplay), preparatory to the return journey outward into the world of event and action. See also Kirby Farrell on "Identity as Magic and Play" in his *Shakespeare's Creation* (Amherst: University of Massachusetts Press, 1975), pp. 178–82.

32. Even Hamlet's language, as it turns positive, acquiring the lean and functional look of action, submits to this reductive negation. What is lost are soliloquy, meditation, commentary, antic dispositions of meaning, and a plenitude of figuration. (The Graveyard Scene might be said to constitute Hamlet's last soliloquy, or perhaps a transition from soliloquy to dialogue, since it is a sustained musing on Death cast in the ostensible form of a dialogue with Horatio and interspersed with actual dialogues with the Gravemaker.) In their place are the workmanlike account of Hamlet's actions at sea, the terse ridicule of Osric, the barrenness of the "readiness" passage, and, at the end of the verbal line, the silences of the duel itself, punctuated by the staccato truncations of the contest—"One . . . No . . . Judgment . . . Have at you now! . . . Nay, come again." Then finally, of course, after a brief recapture of eloquence in Hamlet's dying words, a terminal silence.

33. Of course this self-assigned and unperformable task is not an unmixed curse. For just as he has been a timid suitor to the selves he might become, courting all but choosing none, so Hamlet prefers the freedom of addressing in his imagination a multitude of human vexations—tenting the deepest wounds in nature, life, and himself—as against the constraint of focusing upon a single finite act of revenge. See in this connection Kirby Farrell (see note 31), pp. 178–82.

34. I mean in an aesthetic rather than a moral sense. I quite acknowledge the force of Michael Cameron Andrews' claims that at a level beneath cultural codes and religious prescripts "audiences find the drama of 'one character who has done something and one who is going to get him because he has done it' deeply *satisfying*" ("*Hamlet:* Revenge and the Critical Mirror" [Winter 1978], 8(1):14). But the "getting" may be accomplished in many ways, and Hamlet's failure to accomplish it with some measure of efficiency leads to wholesale carnage.

35. See Sigurd Burckhardt on "How Not to Murder Caesar" (note 12 above), pp. 3–21.

PART III. ENDS AND MEANS: ARTISTIC ARREST AND FUNCTIONAL ERASURE

1. Maurice Charney, *Style in Hamlet* (Princeton: Princeton University Press, 1969), p. 52.

2. Of course the King, as I labored to show in Part I, is never called "Claudius" on stage. We know him by that name from the list of *dramatis personae*, which reflects his identification as "Claudius" in the stage direction preceding Act 1, Scene 2. Assuming that Shakespeare is responsible for that identification—and who else could be?—we can infer that for him the King, whether named or not, was "Claudius."

3. John W. Draper, *The Hamlet of Shakespeare's Audience* (Durham, N.C.: University of North Carolina Press, 1939), pp. 26–27.

4. G. Wilson Knight, *The Wheel of Fire*, 5th rev. ed. (New York: Meridian Books, 1965), p. 38.

5. William Empson, "Hamlet When New," *The Sewanee Review* (1953), 41:15–42; reprinted in *Discussions of Hamlet*, ed. J.C. Levenson (Boston: D.C. Heath, 1960), p. 101.

6. This relationship between function and personality is not restricted to persons. One might point to the bleaching out of personality from figurative speech as it declines into clichés that serve merely as unexamined units of meaning in sentences. Again, what structural grammarians call "function words"—the prepositions, conjunctions, determiners, etc. that serve as syntactic links between the major lexical forms (nouns, verbs, adjectives, adverbs)—are, one might say, devoid of personality in the sense of lexical meaning.

I should note that Terence Eagleton, in a very perceptive chapter on *Hamlet* in his *Shakespeare and Society* (New York: Schocken Books, 1967), singles out "Agency [as] a central theme in *Hamlet*: society is presented, and is present to each character, as a continuous network of causes, agents, and effects, a network of men reciprocally using and exploiting one another. The reciprocity is important: men both use others and are used, and as a result they have two senses of themselves, as source and as agent (or object)" (p. 40). His terms "source" and "agent" correspond to my "personality" and "function."

7. See the chapter called "Recursive Structures and Processes," pp. 127–52 in *Gödel, Escher, Bach: An Eternal Golden Braid* by Douglas P. Hofstadter (New York: Basic Books, 1979).

8. See Terence Eagleton (note 6), p. 41.

9. These distinctions are implicit in Saussure's *Course in General Linguistics*, trans. Wade Baskin (London: McGraw-Hill, 1974) and in the "glossematics" of Hjelmslev—see his *Prolegomena to a Theory of Language*, trans. Francis J. Whitfield (Madison, Wisconsin and London: The University of Wisconsin Press, 1969), first pub. in Copenhagen, 1943—but they are most explicit in Roman Jakobson's famous essay "Two Aspects of Language and Two Types of Aphasic Disturbances" in *Fundamentals of Language* by Jakobson and Morris Halle (The Hague: Mouton, 1956).

10. A more complicated version of this view of literary works is developed by Tzvetan Todorov in his "grammatical" analysis of the stories in the *Decameron*—*Grammaire du Decameron* (The Hague: Mouton, 1969). See also his *La Poetique de la Prose* (Paris, 1971), translated by Richard Howard as *The Poetics of Prose* (Ithaca, N.Y.: Cornell University Press, 1977), esp. "The Grammar of Narrative," pp. 108–19.

11. Roman Jakobson and Morris Halle (see note 9), p. 96. Jakobson's claim that

metaphor and metonymy represent the fundamental paradigmatic/syntagmatic opposition of language has been disputed by Stephen Ullman, who finds the two tropes intimately connected in Proust's imagery (*Style in the French Novel* [Oxford: Blackwell, 1964], pp. 196–207); by Gérard Genette, who finds many Proustian metaphors originating in metonymy ("Métonymie chez Proust," *Figures III* [Paris: Sevil, 1972], pp. 42–43); and by Paul de Man, who relegates key Proustian metaphors to a position subordinate and contributory to metonymy (*Allegories of Reading* [New Haven: Yale University Press, 1979], pp. 65–66). For a discussion of these matters, see Jonathan Culler's "The Turns of Metaphor" in his *The Pursuit of Signs* (Ithaca, N.Y.: Cornell University Press, 1981), pp. 188–209. It should be noted, however, that Jakobson himself admitted that "In poetry, where similarity is superinduced upon contiguity, any metonymy is slightly metaphoric and any metaphor has a metonymical tint" ("Closing Statement: linguistics and poetics" in Thomas A. Sebeok, ed., *Style in Language* [Cambridge, Mass.: MIT Press, 1960], p. 370). One could also contest Jakobson's use of the term "metaphor" to represent association-by-similarity, since metaphor is not the norm but the extreme of similarity, the point at which the differences between vehicle and tenor are greater than their similarities. I use Jakobson's terms here, however, because of their familiarity.

12. T. McAlindon has some excellent observations about the disruption of ceremony in the play as a result of "Gertrude's original sin"—in *Shakespeare and Decorum* (London: Barnes and Noble, 1973), pp. 72–76.

13. Thomas P. Van Laan, *The Idiom of Drama* (Ithaca and London: Cornell University Press, 1970), p. 196. Van Laan's perceptive analysis originally appeared as "Ironic Reversal in *Hamlet*," *SEL* (1966), 6:247–62.

14. Thomas P. Van Laan (see note 13 above), p. 203.

15. See Harold Fisch, *Hamlet and the Word* (New York: Frederick Ungar, 1971), p. 65. Fisch, who registers his indebtedness to an unpublished paper by Hunter Kellenberger entitled " 'Consummation' or 'Consumation' in Shakespeare," stresses the secondary meaning (that is, secondary to the primary meaning of "annihilation") of "satisfaction, fulfillment, joy, even ecstasy" (p. 65), whereas I would give priority, as I think Shakespeare does throughout his plays, to the sense of "completion, of consummating an intention or act."

16. This sentence raises the issue of the differences between speech and writing maintained by Jacques Derrida, who says that speech has at least an "apparent transcendence" and an "immediate presence" because "the phenomenological 'body' of the signifier seems to fade away" in the act of being voiced, thereby affording us immediate access to the signified (*Speech and Phenomena*, trans. David B. Allison [Evanston, Ill.: Northwestern University Press, 1973], p. 77). Moreover, this "self-presence" of the signified is enhanced by the fact that, unlike the writer, the speaker hears his own utterances and hence can modify them to insure that his signifieds are properly manifested. Presumably Derrida would add that the speaker also can gather from his listener's reactions how well he has been understood and can make the appropriate clarifications. For such reasons, speech is relatively risk-free in the communication of its meanings; it does not, in Derrida's view, submit itself to the vicissitudes of the outside world as writing does when it is released onto the page and sent its uncertain way toward a readership, sans author.

However, it should be observed that dramatic speech differs quite radically in this respect from ordinary speech, inasmuch as the playwright's language has been released as a script into the keeping of actors who, in performance, cannot (or can only very slightly) modify what they hear themselves saying. Their words are released from their lips as unalterably as the page proofs of the writer. In fact, Shakespeare gives expression to this fact of theater at an appropriate moment, just before the Inner Play is performed:

King. How fares our cousin Hamlet?
Ham. Excellent, i' faith, of the chameleon's dish; I eat the air, promise-crammed. You cannot feed capons so.
King. I have nothing with this answer, Hamlet. These words are not mine.
Ham. No, nor mine now. (3.2.91–96)

Once Hamlet has uttered his enigmatic puns, they are no longer his; their fate—to be understood or to be rejected as mad-speech—lies with their listeners, both Claudius and the audience in the Globe.

In addition, although its spokenness may lend it an apparent transcendence, its signifieds seeming to be less physically encumbered by its signifiers than is the case with written signs, dramatic speech in Shakespeare contradicts this appearance by virtue of its ubiquitous wordplay, which restores body to the signifier and renders access to the signified indirect indeed.

17. Jacques Derrida (see note 16), pp. 129–60. Actually, there is a significant difference between what Derrida says and what Hamlet does. For Derrida, as for Saussure, the sign itself is "not"—that is, is devoid of positive value, presence, being; it is merely a function within a system of differential relations. But in *Hamlet* the hero's differentiations of himself from others constitute a foregrounding of self, of personal presence; his delay takes place at the cost of function and in behalf of self-definition. Thus the middle of the play does not erase the hero but substantializes him—in part because, as the negative affirms while denying, so *Hamlet is* what he claims not to be. Moreover, he takes on enormous presence in the very process of disclaiming identities; his nay-saying has a positive flair and style that register indelibly in our imaginations.

18. William Empson, "*Hamlet* When New," *Sewanee Review* 1953), vol. 41, reprinted in *Discussions of Hamlet,* ed. J.C. Levenson (Boston: D.C. Heath, 1960), p. 99. Empson believes Shakespeare foregrounded Hamlet's delay as a means of preventing the audience from laughing at what Empson dubiously takes to be the antiquated revenge play structure.

Of course *Hamlet* is not alone in publicizing the presence of delay in its structure. In the Choruses of *The Spanish Tragedy,* for instance, the Ghost of Don Andrea keeps asking Revenge, who at one point falls into a symbolic sleep, why it has been summoned from Hell only to be kept waiting. Here, the Ghost of Don Andrea speaks for the audience's sense of frustration, and Revenge offers sympathy and reassurance on behalf of the playwright, saying in effect, "Yes, it is frustrating at this point, but have patience—relief is at least distantly in sight, and the promised end will rescue all." The middle of the play is thus presented as merely a temporarily annoying means to an exonerating end, and delay is presented, not as an arbitrary given in the generic structure, but as a product of the blocking forces in the hero's society. In *Hamlet,* however, delay is advertised, not to excuse it or to generate suspense, but to call attention to delay as such, to make it a subject of dramatic focus. The phenomenon of delay, of not-action, of juncture is opened to exploration. Thus as Robert Hapgood has shown, delay is not restricted to Hamlet but is a basic and ubiquitous constituent of the dramaturgy of the play—"*Hamlet* Nearly Absurd: The Dramaturgy of Delay," *Tulane Drama Review* (Summer 1965), pp. 132–45.

19. In this respect Shakespeare's naming of delay acts upon his audience the way Prince Hal's "I know you all" soliloquy does in *1 Henry IV* (Act 1, Scene 2). That is, Hal's announcement to the audience of his intention to reform himself later on allays whatever concerns they may have about the political and moral dangers to the state of rampant Falstaff-ism and thus enables them to enjoy the holiday interim secure in the knowledge that a form of comic retribution will come, that Falstaff will not be allowed to spread his girth into the serious affairs of King Henry the Fifth.

20. The comparison of the pausing Pyrrhus to the pausing Hamlet in the Prayer Scene was apparently first made by Harry Levin in *The Question of Hamlet* (New York: Viking Press, 1959), p. 151.

21. Murray Krieger, "The Ekphrastic Principle and the Still Movement of Poetry; or *Laokoön* Revisited," in his *The Play and Place of Criticism* (Baltimore: Johns Hopkins University Press, 1967), pp. 105–28.

22. Bernard Beckerman, *Shakespeare at the Globe: 1599–1609* (New York: Macmillan, 1962), p. 50.

23. Bernard Beckerman (see note 22), p. 47.

24. Mark Rose, *Shakespearean Design* (Cambridge, Mass.: Harvard University Press, 1972), p. 26. See his chapter 'Contexts of Design" for an illuminating account of the Renaissance interest in spatial as opposed to strictly narrative form. Perhaps we should add that the fact that Shakespeare is dealing in *Hamlet* with a conventional causal structure—that is, with an initial cause (the crime) leading to a terminal effect (revenge)—enables him to abandon causation during the episodic middle, rather as the fixed boundaries of a game allow for free play within them.

25. See in this connection Robert Hapgood's "*Hamlet* Nearly Absurd: The Dramaturgy of Delay," *Tulane Drama Review* (Summer 1965), pp. 132–45; and Michael Goldman's chapter "Hamlet and Our Problems" in his *Shakespeare and the Energies of Drama* (Princeton: Princeton University Press, 1972), pp. 74–93. Goldman perceptively discusses how "stop actions" compromise in various ways our expectations of a completed action and thus call in question the very nature of action.

26. Francis Berry, *The Shakespearean Inset* (New York: Theatre Arts Books, 1960), p. 3. Berry's book is an insightful and original study that deserves more recognition than it has received.

27. Francis Berry (see note 26), p. 143. I think I would reverse directions here and argue that the Insets are less the product than the cause of actional impediment in *Hamlet*.

28. Harold Fisch (see note 15), p. 58.

29. This would seem to exemplify Jakobson's famous claim that the "poetic function projects the principle of equivalence from the axis of selection [the paradigmatic] into the axis of combination [the syntagmatic]"—see "Closing statement: linguistics and poetics" in *Style in Language*, ed. Thomas A. Sebeok (Cambridge, Mass.: MIT Press, 1960), p. 358.

30. The interruptions in the first soliloquy are of at least four types, each creating a parenthetical moment in which to comment on the speech itself, either emotionally ("O God, God"), factually ("Nay, not so much, not two"), morally (O, most wicked speed . . ."), or self-reflectively ("Must I remember? . . . Let me not think on 't"). The second

soliloquy (1.5.93–113) is less self-reflective, perhaps because as an immediate response to the harrowing tale and parental demands of the Ghost its focus is inevitably outward, toward the Ghost ("Remember thee!"), toward Gertrude ("O most pernicious woman!"), and toward Claudius ("O villain, villain, smiling, damned villain!"). It is a product of a "distracted globe"—interruptive, skittish, even epileptic in its sudden seizures and dislocations of thought. Apart from the one seven-line sentence in the middle (1.5.99–105), it lacks sufficient coherence and direction to enable interruptive junctures to appear, although there is a momentary self-reflexive fumbling in "What else?/ And shall I couple hell? O, fie!" and a frantic affirmation of his own declarations in "Yes, by heaven!" But the third soliloquy is a clearer example. It is marked by interjections and questions—"For Hecuba! . . . Am I a coward? . . . Who does me this?"—that culminate in a major interruption when he rebukes himself for his loss of verbal control: "Why, what an ass am I!" etc.

31. It is usual to lend a dramatic air to soliloquies and long speeches, as the Bastard does, for instance, in *King John* 1.1.180–219, or as Hamlet himself does in warning his companions not to betray him concerning his "antic disposition" (1.5.169–181). But in Hamlet's soliloquies we have, not the rhetorical imitation of an imagined character by the speaker, but the implicit division of the speaker himself. In converting interior monologue into interior dialogue Hamlet is not so much making a gesture toward the outside world as he is rendering the private utterance self-sufficient by incorporating the "outside" into his own voice as respondent.

32. My somewhat rhetorical personifications of the paradigmatic and the syntagmatic dimensions of language have some basis in their nature and some in the psychology of speakers. The romantic dreaminess I attribute to the paradigmatic—a preference for the imagined over the actual—is owing to the fact that paradigmatic choices are made from a pool of semantic possibilities lying outside the utterance itself, a system of differences that, if examined, ranges far beyond the single term selected. From this paradisal plenitude, by a fatal act of choice, the (let us say) word falls into syntax. But what the paradigmatic construes as a repetition of the primal Fall, the syntagmatic regards unromantically as simply a coming into play (like a player coming off the bench) in order to fill a practical need, the absence of a part. The syntagmatic deals not in virtual relations but in actual relations, not in vertical possibilities but in horizontal realities. It is oriented to function and fabrication, to mortises and plasters and joints and rivets, all that is needed to put together a verbal structure. Clearly the paradigmatic is Shem, the syntagmatic Shaun.

33. William Empson, *Seven Types of Ambiguity*, 2d ed. (New York: Noonday Press, 1955), p. 109.

34. Maurice Merleau-Ponty, *The Prose of the World*, trans. John O'Neill (Evanston, Ill.: Northwestern University Press, 1973), p. 44.

35. Stanley Fish argues that this view of meaning applies to our reading of any sentence and therefore, presumably, of any literary work. Although I disagree with his general claim, I do think it has remarkable relevance to *Hamlet*. See Stanley Fish, *Is There a Text in This Class?* (Cambridge, Mass. and London: Harvard University Press, 1980). In an excellent article "On the Value of *Hamlet*," Stephen Booth makes a similar point without the encumbrances of linguistics:

> Truth is bigger than any one system for knowing it, and *Hamlet* is bigger than any of the frames of reference it inhabits. *Hamlet* allows us to comprehend—hold on to—all

the contradictions it contains. *Hamlet* refuses to cradle its audience's mind in a closed generic framework, or otherwise limit the ideological context of its actions.

(*Reinterpretations of Elizabethan Drama*, ed. Norman Rabkin [New York and London, 1969], p. 175.) I suppose one could also claim that *Hamlet* provides an excellent illustration of Jacques Derrida's concept of the "trace," whereby the not-there of potential meanings imposes its presence upon what is apparently there.

36. P.J. Aldus is, so far as I know, the first to call attention to the plethora of voyages in *Hamlet* and to wonder about their significance—see his *Mousetrap: Structure and Meaning in Hamlet* (Toronto: University of Toronto Press, 1977), esp. the section on "Dream Travellers." See also the interesting article by Keith Brown, "Hamlet's Place on the Map." *SS* (1968), 4:160–82.

37. James Joyce, *Portrait of the Artist as a Young Man* (New York: Compass Books edition, 1956), p. 215.

38. Bridget Gellert Lyons, *Voices of Melancholy* (New York: W.W. Norton, 1971), p. 95.

39. *Johnson on Shakespeare*, ed. Walter Raleigh (London: Oxford University Press, 1908), p. 27.

40. *Johnson on Shakespeare* (see note 39), pp. 161–62; the following quotes are from pages 193 and 196.

41. Eric Bentley, *The Life of the Drama* (New York and London: Athenaeum and Methuen, 1964), p. 214.

42. In regard to the corporealization of language in poetry, see Sigurd Burckhardt, "The Poet as Fool and Priest," in his *Shakespearean Meanings* (Princeton: Princeton University Press, 1968), pp. 22–46. For a more extensive argument in favor of a poetics of presence, see the last two chapters of Murray Krieger's *Theory of Criticism* (Baltimore and London: Johns Hopkins University Press, 1976), pp. 179–245.

43. Our imaginative entrance into the fictional world is facilitated, as it is not in other literary forms, by the communal nature of the theatrical experience, by the fact that we are surrounded by others whose involvement in the fiction reinforces and guides our own. This imaginative engagement was easier for an audience in Shakespeare's time than for one in ours because our awareness of the long history of Shakespearean productions enables us more readily to distinguish the play (whatever that is!) from this particular performance of it, and the character of Hamlet (whoever he is!) from this particular actor who plays him. In Shakespeare's day the audience knew well enough that Hamlet was part of Burbage's repertoire of roles, but they did not yet know that Burbage was part of Hamlet's repertoire of actors. Thus the Elizabethan audience could more easily be led to assume that for the duration of the performance Burbage was Hamlet and the Globe theater was Elsinore. But of course that is precisely what Shakespeare will not allow his audience to assume for long.

44. It is of course arguable that Hamlet finds his way out of the Sleugh of Despond rather too easily. Thus Sigurd Burckhardt says, "In the end he settles for a stoic resignation which is moving and impressive, but which nevertheless signifies an abdication from his task to discover a new order" (see note 42), p. 15. Perhaps the answer to this is that Hamlet's task is not to discover a new order but simply to kill the King. So thoroughly stewed in

corruption is the old order of Claudius, not to mention Hamlet himself, that it is a major accomplishment merely to purge the state by striking at the source of its evils. Hamlet's task is destructive, not creative, an act of erasure that leaves Denmark under Fortinbras free to fashion a new order, or even a new disorder, or, more likely, to recover something of the lots order of Old Hamlet's earlier days.

45. Of course all fictions are imagistic-verbal presences containing, like figurative language, an implicit negative, a conceptual "not this." To imagine, as Sartre emphasized, is to address oneself to what is absent, to create images of what is "not." In this sense literature gives being to not-being. In its crudest form, as direct mimesis, literature harks back to the absent subject and derives its power and value from it. In more sophisticated form, as in *Hamlet*, literature refers not to an absent, already-formed subject that it seeks to reproduce, but to the more deeply absent potentialities of life that find their form in the work itself.

46. The reference of course is to the sign of the Globe theater, which was an image of Hercules, in relief of Atlas, holding up the globe.

47. Shakespeare's having it both ways in *Hamlet* is typical of fictionists, whose imaginative constructions both are and are not. As Paul Ricoeur says, "The poet is this genius who generates split references *by* creating fictions. It is in fiction that the 'absence' proper to the power of suspending what we call 'reality' in ordinary language concretely coalesces and fuses with the *positive insight* into the potentialities of our being in the world which our everyday transactions with manipulatable objects tend to conceal" ("The Metaphoric Process as Cognition, Imagination, and Feeling" in *On Metaphor*, ed. Sheldon Sacks [Chicago: University of Chicago Press, 1978], p. 153). It is not, in other words, that Shakespeare is doing something entirely new—even Majorcan storytellers begin with the expression "it was and it was not"—but that in *Hamlet* he ingeniously complicates and subtilizes the issue.

48. Harold Toliver, *The Past That Poets Make* (Cambridge, Mass. and London: Harvard University Press, 1981), pp. 19–21. Toliver is speaking not so much of imaginative digressions within works as of literature itself as imaginative digressions from the world of ongoing daily affairs—digressions whose absorption in their own innovations may risk passing beyond recognizable relevance into a kind of mad pointlessness.

49. That is, the parallel between the madness of digression and that of Hamlet's postponed revenge is reinforced if we observe that Hamlet's major geographical digression from his revenge, when he is shipped to England, is a product of his "madness." Gertrude has just told the King that Hamlet, "Mad as the sea and wind, when both contend/ Which is the mightier" (4.1.7–8), has killed Polonius, after which the King says, "The sun no sooner shall the mountains touch/ But we will ship him hence" (4.1.29–30). The imagination that wanders truly into madness passes beyond the boundaries of received thought and cannot return to tell its visions; on those wild seas are no benevolent pirates. But the "mad" Hamlet escapes from his parenthesis of not-revenge to perform a meaningful act in the world; and Shakespeare, "mad in craft" (3.4.195) emerges from the digressive parenthesis of the middle of his play to connect it to its generic world by means of revenge.

50. Jan Kott, *Shakespeare, Our Contemporary* (New York: Doubleday, 1964), p. 65.

51. Francis Fergusson, *The Idea of a Theater* (Princeton: Princeton University Press, 1949), p. 122; see also pp. 151–52.

52. "Voice" means of course "vote," but it also means "utterance" or "breath," as it more obviously does a bit later when Horatio, referring to Hamlet, says "And from his mouth whose voice will draw on more" (5.2.394). Fortinbras, I might add, is related to Hamlet rather as Hotspur is related to Prince Hal. Both plays feature a triadic pattern in which the hero is flanked by two characters whose virtues and vices he incorporates either more (Hal) or less (Hamlet) successfully. Norman Holland constructs a full-scale interpretation of *Hamlet* along these lines, with Hamlet divided between Fortinbras the soldier and Horatio the philosopher, and suggests a rather neat emblematic conclusion to the play in which "the man of thought and the man of action stand, facing each other, separated, divided; between them lies the man who was both, a failure precisely because he was both" (*The Shakespearean Imagination* [New York and Toronto: Macmillan, 1964], pp. 178–79).

53. Another parallel between the two princes is their receiving "commissions" from their uncle-kings. Old Norway gives Fortinbras "his commission to employ these soldiers,/ So levied as before, against the Polack" (2.274–75) as a reward for abandoning his revengeful war against Denmark. By acting in accord with the terms of this commission Fortinbras makes his individual mark and serves his father's cause as well. Hamlet, on the other hand, receives his commission as a penalty for pursuing his revengeful cause, especially for killing Polonius. By rejecting it and rewriting its terms, he too makes his individual mark and serves his father's cause. K.R. Eissler presents an extended discussion of "Fortinbras and Hamlet" in his *Discourse on Hamlet and Hamlet: A Psychoanalytic Inquiry* (New York: International Universities, 1971), pp. 316–40, in which he suggests that Fortinbras is a kind of "Hamlet *redivivus*" (p. 324).

54. Ernst Kantorowicz, *The King's Two Bodies* (Princeton: Princeton University Press, 1957).

55. From this point of view there is a curious presence/absence about the reign of Claudius, who is indispensably present as regards the accession of Fortinbras but merely a self-erasing parenthesis between "Hamlet" and "Hamlet." With Fortinbras now symbolically equated with "Hamlet," Claudius seems both to be and not to be: "The King is a thing— . . . Of nothing" (4.2.29 ff.). If, on the other hand, Laertes had been pressed into kingship by the mob—he who, unlike Fortinbras, has no parallel to Hamlet in the royal succession—then the reign of Claudius would have been entirely transitive and fully present, and the Danish succession would have become a degraded linear process devoid of cyclical overtones.

56. It should be added that the return of Fortinbras from the Polish wars also reflects Horatio's mention of Old Hamlet's action against the Poles in the opening scene (1.1.62–63). In this connection Francis Berry argues that Horatio's description of the dead king as a "martial hero if ever there was one," followed by a similar depiction of the warlike Fortinbras (1.1.95–104), makes it "difficult not to feel that the spiritual, if not the fleshly, son of Hamlet I is not Fortinbras II" and therefore that the "dying voice" of Hamlet falls on Fortinbras with peculiar appropriateness (see note 26), pp. 124–25. Maurice Charney observes in a similar vein, "The prominence of the Ghost in the first act is matched by the emergence of Fortinbras at the end of the play as the new king of the Danes, so that both the beginning and the end of *Hamlet* have a strong military interest" (see note 1), p. 28.

57. *King Lear,* 5.3.325–326. Walter C. Foreman has a perceptive discussion of the relations between the "old order" and the "new order" at the end of Shakespeare's tragedies, in *The Music of the Close* (Lexington, Ky.: University of Kentucky Press, 1979), pp. 1–28.

58. It is equally true, of course, that the visual features of the play are also left behind as the play moves on. In *Hamlet,* however, especially at the end, it is the transiency of what is heard that Shakespeare emphasizes (in contrast, say, to *The Tempest,* where his stress is on the transiency of what is seen—e.g., the "baseless fabric of this vision," the vanishing "cloud-capped towers," etc. [4.1.151 ff.]).

59. Geoffrey Hartman, *Beyond Formalism* (New Haven: Yale University Press, 1970), p. 341. Some linguists regard juncture not merely as a breathing space but as a combination of pitch, stress, and pause; others regard it as a matter of pitch and pause; and still others as a matter of pause only. Obviously I am more interested here in the implications of Shakespeare's metaphors than in exact descriptions of suprasegmental phonemes.

60. To pursue this line with both bated and baited breath, we might suggest that the overall rhythm of the play is that of breathing, a pattern of self-relinquishing exhalations and self-recovering inhalations associated with diachronic flow and synchronic arrest. I suppose on this view that Hamlet's delay would have to be regarded as a sustained holding of his breath, which in its own way would be as life-threatening as an inability to catch his breath because of ceaseless activity. Perhaps that has a certain symbolic rightness about it: Hamlet does in a sense refuse to breathe the air of the world he finds himself in for fear it will pollute him; and it is only when he has sensed a divine ventilation of Denmark after his ocean voyage that he is able to breathe easily. Clearly this airy subject calls for total oxygenation in a doctoral thesis.

EPILOGUE: PRESENCE AND ABSENCE

1. Ferdinand de Saussure, *Cours de linguistique générale,* 3d ed. (Paris: Payot, 1967); trans. *Course in General Linguistics,* Wade Baskin (New York: McGraw-Hill, 1966). Jacques Derrida, *L'Ecriture et la différence* (Paris: Seuil, 1967); trans. *Writing and Difference,* Alan Bass (Chicago: University of Chicago Press, 1978). There is of course an ever-growing library of works dealing with the relations between linguistics, Russian formalism, structuralism, and post-structuralism; see the excellent bibliography in Terence Hawkes' *Structuralism and Semiotics* (London: Methuen, 1977). See also the excellent analysis of poetic presence and absence in the final chapter of Murray Krieger's *Theory of Criticism* (Baltimore and London: Johns Hopkins University Press, 1976), pp. 207–45. Krieger's collection of essays, *Poetic Presence and Illusion* (Baltimore and London: Johns Hopkins University Press, 1979), is also continuously relevant. See also, in *Deconstruction & Criticism* (New York: Seabury, 1979) the various essays by Harold Bloom, Paul de Man, Jacques Derrida, Geoffrey Hartman, and J. Hillis Miller. Also the chapter "Criticism, Indeterminacy, Irony" in Hartman's *Criticism in the Wilderness* (New Haven: Yale University Press, 1980), pp. 265–83. Finally, the fourth and fifth chapters of Frank Lentricchia's *After the New Criticism* (Chicago: University of Chicago Press, 1980), pp. 102–210, present an excellent discussion of structuralism and post-structuralism.

2. This is the translation of Jonathan Culler in *Structuralist Poetics* (London: Rout-

ledge and Kegan Paul, 1975), p. 247, based on Derrida's remark in *L'Ecriture et la différence* (Paris: Seuil, 1967), p. 411. Alan Bass translates the remark as "The absence of the transcendental signified extends the domain and the play of signification infinitely" (*Writing and Difference* [Chicago: University of Chicago Press, 1978], p. 280).

3. Julia Kristeva, *Semiotikè: Recherches pour une semanalyse* (Paris: Seuil, 1969), p. 284. Kristeva's geno-text, with its notion of infinite verbal possibilities underlying the literary text, carries the laudable concept of polysemous meaning in literature to the self-defeating point at which no meaning can be rejected. One can applaud Kristeva's desire to alert readers to the full range of meaning in language, especially in literary language, and yet maintain that the very concept of meaning implies than an enormous number of verbal possibilities are excluded in any expression.

4. As Murray Krieger observes, "literature itself is no enemy to the deconstructive impulse. Far from it. Indeed, one might well argue that in its reflexivity and self-consciousness literature not only deconstructs itself but is the very model for our use in the deconstruction of other discourse" (see note 1), p. 181. Similarly, J. Hillis Miller argues that great literary works have anticipated the deconstructive efforts of critics and that their indeterminacy of meaning "is always thematized in the text itself in the form of metalinguistic statements" ("Deconstructing the Deconstructors," *Diacritics* [Summer 1975], 5[2]:30–31).

5. One might say that Hamlet has recognized an element of metonymic difference within metaphoric likeness. Metonymic contiguity offers him a close but independent difference from his father in a way that metaphoric identity does not. Or one might simply say that Hamlet is keeping the metaphor metaphoric by maintaining a certain tension and distance between its tenor and vehicle, a sense of difference that is essential to living metaphors and altogether lost in "faded" or "dead" metaphors, which are no longer metaphors at all but literal identities. Had Hamlet succeeded his father on the throne or had he totally submerged himself in the role of revenger, he would have killed the metaphor that sustains the life of his independent self.

6. That is, if we think of the radical particular as constituting one extreme and the concept or universal as the other, the former representing complete unlikeness and the latter complete likeness, then metaphor lies somewhere in between, asserting complete likeness but smuggling into that assertion an implicit acknowledgement of difference.

7. And actions in themselves, we should remember, have no negative; like physical objects, they are simply there, neither true nor false. Negation, however, may be suggested by a manipulation of the context of action, as the scene of a boxing match or of a play suggests that these actions are not what they appear to be, real fighting or real behavior. Yet the actions in themselves are entirely positive and fully present.

Index